SPECIAL PAPERS IN PALAEONTOLOGY NO

CAMBRIAN BRADORIID AND PHOSPHATOCOPID ARTHROPODS OF NORTH AMERICA

BY

DAVID J. SIVETER and MARK WILLIAMS

with 9 plates and 8 text-figures

Special Papers in Palaeontology 57
This issue has been reprinted completely due to an error in production at the printers & is intended to replace the original copy

THE PALAEONTOLOGICAL ASSOCIATION
LONDON
SEPTEMBER 1997

Printed in Great Britain

CONTENTS

ABSTRACT. North American species were among the first bradoriid and phosphatocopid arthropods to be documented and include many historically and nomenclatorially important taxa.

The most abundant specimens of bradoriids and phosphatocopids in North America occur in the Canadian Atlantic Provinces, British Columbia and Arizona and, to a lesser extent, in New York State and Tennessee. The families Bradoriidae (junior synonym: Indianidae), Beyrichonidae, Hipponicharionidae, Svealutidae and Vestrogothiidae are represented; *Escasona* and the hitherto widely used *Aluta* are considered to be *nomina dubia*. The total known fauna, comprising 16 genera (including *Matthoria* gen. nov.) and 26 species (including three under open nomenclature and *Liangshanella burgessensis* sp. nov.), is much less diverse at the specific level than indicated in previous records (about 100 species). The genera *Beyrichona*, *Hipponicharion*, *Liangshanella* and others have a phosphatic carapace, thus endorsing the notion that carapace composition cannot be used as a diagnostic criterion for distinguishing bradoriids from phosphatocopids.

Provinciality of the bradoriid and phosphatocopid faunas supports the notion of an Iapetus Ocean in the Cambrian. Eastern Canadian (Avalonian microcontinent) faunas are characterized by *Beyrichona*, *Cyclotron* and *Hipponicharion*; Arizona (Laurentian palaeocontinent) has *Walcottella* and *Dielymella*. *Bradoria* is known from both the Avalonian and Laurentian parts of North America; the cosmopolitan *Anabarochilina* occurs only in the Laurentian part. Faunas of the distal shelf of Laurentia have Asiatic (*Liangshanella* and cambriids) and Australian (*Indota*) bradoriids.

Bradoriids and phosphatocopids occur throughout the Cambrian of both the Avalonian and Laurentian parts of North America. Most species are short ranging but have only local geographical occurrence. A few species have intercontinental biostratigraphical value, being coeval in Canada, Britain and Scandinavia.

The bradoriid and phosphatocopid faunas of North America occur as low diversity marine assemblages and were probably mostly benthic or nekto-benthic. Shallow water cratonic sequences in Arizona and distal shelf/shelf edge deposits in Alberta, British Columbia and in Greenland (all on palaeocontinental Laurentia) have very low diversity bradoriid assemblages (one or two species). Shallow shelf sequences in New Brunswick and Nova Scotia (palaeocontinental Avalonia) contain mainly richer, but also low diversity bradoriid assemblages (three to six species). Distal shelf faunas in Newfoundland have monospecific phosphatocopid assemblages.

BRADORIID and phosphatocopid arthropods typically have small (length *c.* 1–17·5 mm), bivalved carapaces, are almost exclusively Cambrian in age and have conventionally been referred to the ostracod crustaceans. However, the bradoriid and phosphatocopid carapace, which differs from that of ostracods in lacking an articulating hinge structure and, in some cases, in having a calcium phosphate composition, is not a conclusive indicator of affinity. Indeed, instances of exceptional soft-part preservation indicate that phosphatocopids in general, and at least some bradoriids, should be regarded as Crustacea *s.l.* rather than ostracods (e.g. Müller and Walossek 1991; Walossek and Müller 1992; Hou *et al.* 1996).

Bradoriids and phosphatocopids had world-wide distribution and formed a major faunal element of the arthropod fauna from the early Cambrian until the earliest Ordovician (see Huo and Shu 1985; Huo *et al.* 1991; Siveter *et al.* 1996 and references therein). For example, in the early Cambrian Chengjiang *Lagerstätte* of China, bradoriids account for some 80 per cent. of recovered individuals (Hou and Bergström 1991). Recent studies have demonstrated the biostratigraphical and biogeographical potential of bradoriids and phosphatcopids (e.g. Williams *et al.* 1994*c*; Siveter and Williams 1995; Siveter *et al.* 1996; Melnikova *et al.* in press).

Occurrences of the Cambrian bradoriid and phosphatocopid faunas of North America were reported originally during the late nineteenth century. These were among the first such assemblages to be documented and therefore (together with British faunas: see Rushton *et al.* in press; Williams and Siveter in press) include many historically and nomenclatorially important taxa. The early study of bradoriids and phosphatcopids owes much to the detailed investigations of George Frederic Matthew who, partly under the auspices of the Canadian Geological Survey, undertook fundamental primary studies of the geology and palaeontology of New Brunswick and Nova Scotia. Matthew's most important papers, on what were then regarded simply as ostracods and associates, were published between 1886 and 1906 (see references herein). Miller and Buhay (1990) provided an account of Matthew's life and work and a bibliography of his publications.

Although Walcott (e.g. 1887, 1890) published only a few descriptions of 'bradoriids' *s.l.*, he made extensive collections of the group in eastern Canada, British Columbia and northern Arizona, a task which culminated in Ulrich and Bassler's (1931*a*) monograph of the entire North American 'bradoriid' fauna. No revision of Matthew's, Walcott's or Ulrich and Bassler's work has previously been undertaken and since the 1930s only a few additional North American bradoriid and phosphatocopid species have been recorded, mainly from sparse material (e.g. Resser 1938; Frederickson 1946; Palmer 1954; Lochman 1956; Lochman and Hu 1960; Bolton and Copeland 1963; Landing 1980; Martin and Dean 1988). The need to 'rehabilitate' all these North American faunas, especially the nomenclatorially important genera established by Matthew, and to resolve their biogeographical provenance, palaeoenvironmental setting and stratigraphical potential, has been highlighted by recent taxonomic and biostratigraphical investigations of bradoriid and phosphatocopid faunas from the Cambrian of Scandinavia, Australia, China, Britain and the former Soviet Union (e.g. Melnikova 1984, 1987, 1988; Huo *et al.* 1991; Hinz-Schallreuter 1993*a*, 1993*b*; Hinz-Schallreuter and Jones 1994; Williams *et al.* 1994*c*; Siveter and Williams 1995; Siveter *et al.* 1996 and references therein; Melnikova *et al.* in press; Rushton *et al.* in press; Williams and Siveter in press).

The only palaeocontinental Laurentian bradoriid material known from outside North America comprises three species from Greenland (Poulsen 1932; Siveter *et al.* 1996) which, when referring to Laurentian faunas, are also included herein.

PRESERVATION, TECHNIQUES, TERMINOLOGY AND COLLECTIONS

Preservation

Virtually all of the North American bradoriids and phosphatocopids occur as incomplete 'crack-out' valves on slabs of either mudstone, sandstone or limestone. Valves are often flattened or distorted and in some instances the evidence indicates that the carapace may have been originally soft and somewhat flexible. Rarely, some phosphatocopids have been extracted from limestones using acid techniques (e.g. Landing 1980). No preservation of bradoriid or phosphatocopid soft parts is known from North America, even from the bradoriids of the Burgess Shale.

Techniques

Specimens have been photographed mainly by light photography using an Aristophot mounted with a Leica camera (methods of Siveter 1990). Immediately prior to photography, specimens were coated with a thin film of ammonium chloride sublimate to enhance contrast. Small specimens and details of ornament were studied using the scanning electron microscope, and were coated with a removable silver coating (method of Mills 1988). Energy dispersive X-ray analytical techniques (EDX) were used to determine the carapace composition of some non-type specimens.

Terminology

For purely practical purposes, the terminology used for the features of the carapace of post-Cambrian ostracods is herein applied to the Cambrian phosphatocopids and bradoriids (see Siveter *et al.* 1993, 1996; Williams *et al.* 1994*a*, 1994*b*, 1994*c*). The term 'interdorsum' follows the usage of, for example, Hinz (1993).

Collections

Collections of North American bradoriids and phosphatocopids are housed both in North American and in European institutions. The type and other collections of Matthew, from New Brunswick and Nova Scotia, are in the New Brunswick and Royal Ontario museums.

Royal Ontario Museum (ROM), Toronto, Canada. Has type and figured material from Matthew's papers published between 1886 and 1906 (see references). This includes the type species of Matthew's genera *Aluta*, *Beyrichona*, *Bradoria*, *Hipponicharion* and *Indiana*, and *Mononotella* Ulrich and Bassler, 1931*a*. Within syntype suites assigned a single ROM number individual specimens are labelled A, B, C, etc. (see Waddington *et al.* 1978).

New Brunswick Museum (NBMG), St John, Canada. Has Matthew's unfigured type and non-type collections from Nova Scotia and New Brunswick (see Miller 1988). Also houses extensive collections made in New Brunswick and Nova Scotia during August 1994, by DJS, MW and R. F. Miller (NBMG).

Geological Survey of Canada (GSC), Ottawa, Canada. Contains the type material of *Bullaluta* Copeland, 1986 from the Cow Head Peninsula of western Newfoundland and specimens of *Cyclotron* Rushton, 1969 and Rushton's (1978, pl. 24, fig. 9) specimen referred to as *Walcottella* from the Manuels River area of eastern Newfoundland (see Martin and Dean 1988).

United States National Museum Smithsonian Institution (USNM), Washington D.C., USA. Houses an extensive collection of bradoriids and phosphatocopids, largely from New Brunswick, Nova Scotia, the Grand Canyon and the Burgess Shale of British Columbia. Includes the type species of *Dielymella* and *Walcottella* (both Ulrich and Bassler, 1931*a*). Ulrich and Bassler (1931*a*) often assigned a single USNM number to a suite of specimens which they considered to be a single taxon. Because these suites often contain more than one figured specimen and/or specimens of questionable or indeterminate affinity, in many cases it has been necessary to assign new, unique numbers to many of their specimens (USNM 483167–483196, 483882–483891, 486352–486356, 486426–486437, 487774–487782).

University of Kansas Museum of Invertebrate Paleontology (KUMIP), Lawrence, Kansas, USA. Contains material mainly from Utah, collected by R. A. Robison.

University of Texas (UT), Austin, Texas, USA. Contains material from Texas (Palmer 1954).

University of Oklahoma Museum of Invertebrate Paleontology (OUMIP), Norman, Oklahoma, USA. Contains the holotype left valve of '*Leperditia*' *harrisi* Frederickson, 1946.

University Museum Oxford (OUM), Oxford, UK. Houses collections made in New Brunswick, in August 1994, by DJS, MW and R. F. Miller.

The Natural History Museum (NHM), London, UK. Contains material from New York State.

Geological Museum of the University of Copenhagen (MGUH), Denmark. Has collections from North Greenland (Siveter *et al.* 1996).

In addition to the above institutions, specimens relevant to discussion of North American bradoriids and phosphatocopids are housed in the British Geological Survey (BGS), Keyworth, Nottingham, UK; Bristol City Museum and Art Gallery (BRSMG), Bristol, UK; and the Palaeontological Museum (PMU), University of Uppsala, Sweden.

STRATIGRAPHY AND LOCALITIES

Within North America, bradoriids and phosphatocopids occur in rocks of Early, Mid and Late Cambrian age and are geographically widespread (Text-figs 1–2). However, the precision of information about localities and stratigraphy is variable. Some localities, for example those in New

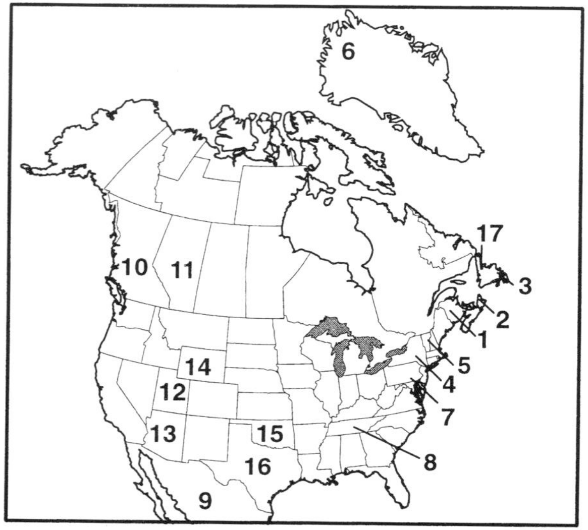

Epoch	Avalonia			Laurentia	
	New Brunswick	Nova Scotia	East New-foundland	New York State	Vermont
Late Cambrian	Chesley Drive Group 1		Maclean Brook Formation 3		
Mid Cambrian	Chamberlains Brook Formation 1	Dugald & Maclean Brook formations 2			
Early Cambrian	Hanford Brook Formation 1			*Elliptocephala asaphoides* fauna 4	Sandstones at Sunset Hill 5

Epoch	Laurentia											
	North Greenland	Penn-sylvania	Tennessee	Mexico	British Columbia	Alberta	Utah	Arizona	Wyoming	Oklahoma	Texas	West New-foundland
Late Cambrian									Du Noir Limestone 14	Honey Creek Formation 15	Riley Formation 16	*Cedaria-Crepicephalus* Zone 5 conglomerate 17
Mid Cambrian					Stephen Formation 10	Shales at Windsor Mountain 11	Marjum and Weeks formations 12	Bright Angel Shale 13				
Early Cambrian	Buen Formation 6	Kinzers Formation 7	Murray Shale 8	Puerto Blanco Formation 9								

TEXT-FIG. 1. Summary of the stratigraphical and geographical distribution of bradoriids and phosphatocopids in the Cambrian of North America and Greenland, incorporating the palaeocontinents of Laurentia and Avalonia.

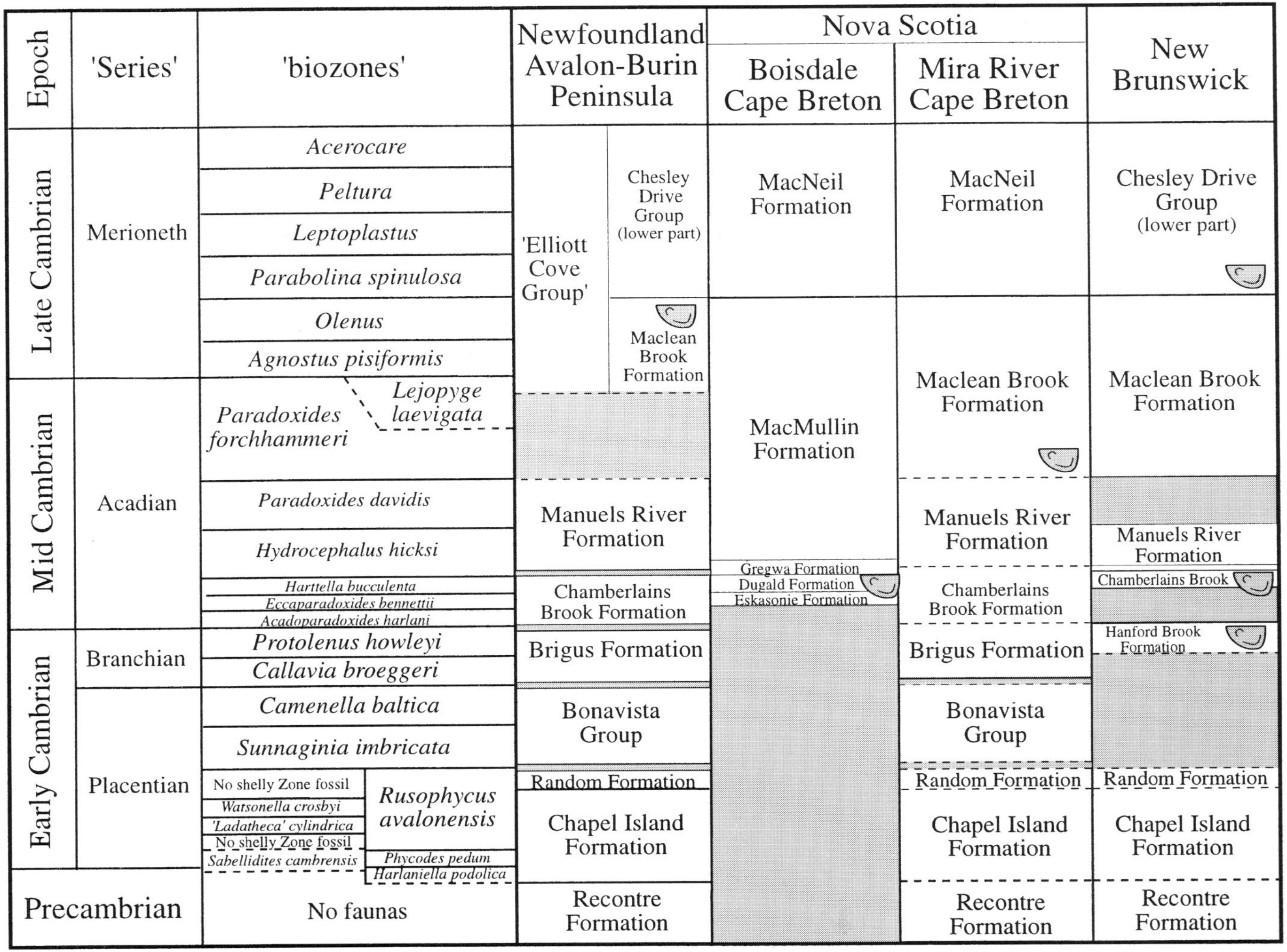

TEXT-FIG. 2. Cambrian stratigraphy and distribution of bradoriids and phosphatocopids in New Brunswick, Nova Scotia and eastern Newfoundland. Stratigraphy summarized from North (1971) Landing, (1980, 1991, 1992, 1995, 1996), Martin and Dean (1988), White *et al.* (1994) and Landing and Westrop (1996).

Lithostratigraphy	Bourinot Group															
	Dugald Formation															
Bed Numbers (after Matthew 1903)	base of Dugald Formation	E1a	E1b	E1c	E1d	E1e	E2a	E2b	E2c	E3a	E3b	E3c	E3d	E3e	E3f	top of Dugald Formation
Measurements (after Matthew 1903)		43 ft (13.11m)	34 ft (10.36m)	21 ft (6.40m)	21 ft (6.40m)	3 ft (0.91m)	73 ft (22.25m)	57 ft (17.37m)	21 ft (6.40m)	54 ft (16.46m)	25 ft (7.62m)	25 ft (7.62m)	30 ft (9.14m)	18 ft (5.49m)	82 ft (24.99m)	

TEXT-FIG. 3. Matthew's (1903) stratigraphical divisions (with thicknesses) for the early Mid Cambrian Dugald Formation of Cape Breton, Nova Scotia.

York State (e.g. Ford 1873; Walcott 1887), have not been recollected for over a century. Most localities have yielded only a few specimens, often in monospecific assemblages, as in Pennsylvania, Oklahoma, Texas and Wyoming (e.g. Resser and Howell 1938; Frederickson 1946; Palmer 1954; Lochman and Hu 1960). More abundant, but not necessarily diverse faunas, are known from British Columbia, the Canadian Atlantic Provinces, northern Arizona, Tennessee and North Greenland (e.g. Matthew 1886, 1903; Ulrich and Bassler 1931*a*; Laurence and Palmer 1963; Siveter *et al.* 1996).

Canadian Atlantic Provinces

In his Cambrian stratigraphy of New Brunswick and Nova Scotia, Matthew (e.g. 1886, 1902, 1903; see Text-fig. 2) used a series of lithologically and palaeontologically based divisions, many of which correspond to more formally defined formations. Where appropriate, we have used Matthew's stratigraphical nomenclature when discussing the distribution of species within the Dugald Formation (Text-fig. 3).

We have made extensive new collections of the faunas originally described by Matthew at two localities, Hanford Brook in New Brunswick and Dugald Brook in Nova Scotia (Text-fig. 4A–B), both of which have yielded several nomenclatorially important taxa including the type species of *Bradoria* itself (see Matthew references 1886–1906).

New Brunswick. These localities are situated within and to the north of the city of St John (Text-fig. 4A) and include:

Hanford Brook (Matthew 1886, 1895; McCutcheon 1987): Loch Lomond sheet 21 H/5, Grid Ref. 960 366. Here the richly fossiliferous Early Cambrian type Hanford Brook Formation (Tanoli and Pickerill 1988; Landing and Westrop 1996) yields the type species of *Beyrichona*, *Hipponicharion* and *Mononotella* (Matthew 1886; Ulrich and Bassler 1931*a*).

Long Island (Matthew 1898): St John sheet 21 G/8, Grid Ref. 320 295. Here the Hanford Brook Formation yields *Indiana* and *Beyrichona*.

Porter Road (Matthew 1896; Hayes and Howell 1937): Loch Lomond sheet 21 H/5, approx. Grid Ref. 942 330. This is the locality for the Mid Cambrian Fossil Brook Member of the Chamberlains Brook Formation (see Landing 1996) which yields the type species of *Aluta*.

Hastings Cove (Ulrich and Bassler 1931*a*): St John sheet 21/G8, Grid Ref. 325 254. Has yielded *Indiana* from Mid Cambrian limestones.

Germaine Street, within the city of St John (see Landing 1980, text-fig. 1). Has yielded phosphatocopids from the Late Cambrian part of the Chesley Drive Group (see Landing 1980, 1996).

Nova Scotia. Localities in the vicinity of Lake Bras d'Or and the Mira River, Cape Breton (Text-fig. 4B), include:

Dugald Brook near the village of Eskasonie, and smaller exposures in Gillis Brook, Indian Brook and Boundary Brook (see Matthew 1899, 1902, 1903): Grand Narrows sheet 11F/15 (see also Hutchinson 1952,

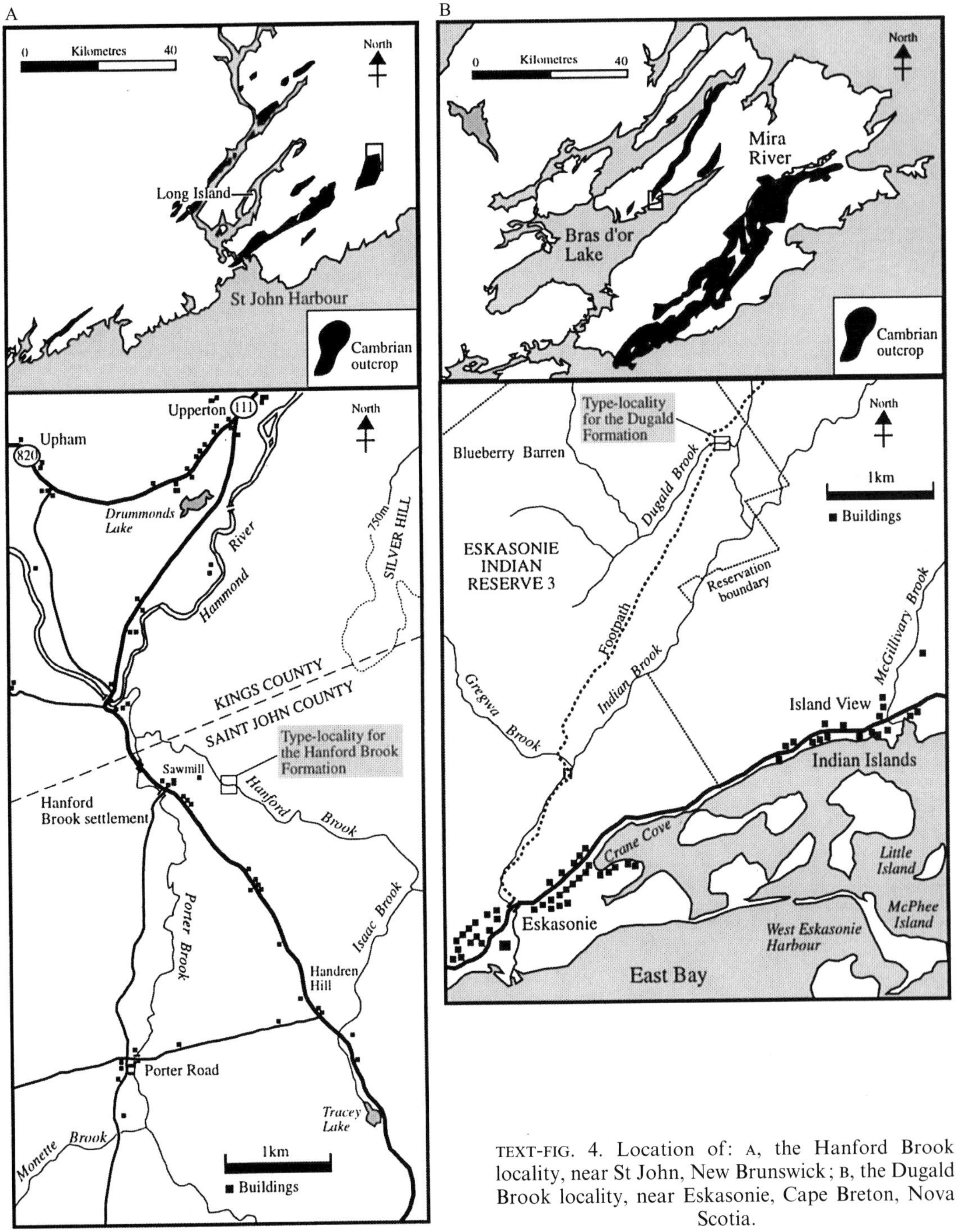

TEXT-FIG. 4. Location of: A, the Hanford Brook locality, near St John, New Brunswick; B, the Dugald Brook locality, near Eskasonie, Cape Breton, Nova Scotia.

fig. 2; White *et al.* 1994, fig. 2). Exposes the Mid Cambrian type Dugald Formation, type locality of *Bradoria* and *Indiana*.

Mira River, near Marion Bridge, *c.* 2·5 km west of Marion Bridge and MacNeil Brook, east of Marion

Bridge (see Matthew 1903; Ulrich and Bassler 1931*a*; Hutchinson 1952; Landing 1991). Here the late Mid Cambrian part of the Maclean Brook Formation yields *Beyrichona*, *Septadella* Stubblefield, 1933 and possible *Wimanicharion* Hinz-Schallreuter, 1993*b*.

Newfoundland

The Upper Cambrian at Manuels River, eastern Newfoundland, yields *Cyclotron* Rushton, 1969 (Martin and Dean 1988, figs 1–2; Williams *et al.* 1994*c*; for revised stratigraphy see Landing 1996).

The Late Cambrian *Cedaria-Crepicephalus* Zone 5 conglomerate cropping out along the north-east shore of the Cow Head Peninsula, western Newfoundland (approx. lat. 49° 55′ 30″ N, long. 57° 48′ W), yields *Bullaluta* Copeland, 1986.

Other North American localities

Lower Cambrian:

The Murray Shale, Chilhowee Group, Tennessee (Resser 1938; Laurence and Palmer 1963).

Limestones from east of the City of Troy, near Greenwich Township, Washington County, and from the Cambridge and Hoosick quadrangles, New York State (see Ford 1873; Walcott 1887, 1890; Ulrich and Bassler 1931*a*; Lochman 1956).

Sandstones from Sunset Hill, near Lake Dunmore, Salisbury, Vermont (Ulrich and Bassler 1931*a*).

The Porto Blanco Formation, Mexico (McMenamin and McMenamin 1990).

Middle Cambrian:

The Burgess Shale at the Raymond and Walcott quarries, near Field, British Columbia (e.g. Conway Morris 1986).

Shales on the west flank and along the north ridge of Windsor Mountain, south-western Alberta (Bolton and Copeland 1963).

The Bright Angel Shale at Bass Canyon, Nunkoweap Valley, Kwagunt Valley and Chuar Valley, Grand Canyon, northern Arizona (Ulrich and Bassler, 1931*a*; see Beus 1987 for stratigraphy): yielded the type material of *Walcottella* and *Dielymella*.

The Howell Formation and uppermost Marjum and Weeks formations of the House Range, Utah (see Hintze and Robison 1975 for stratigraphy).

Upper Cambrian:

The Du Noir Limestone, Sheep Mountain, Wyoming (Lochman and Hu 1960, text-figs 1, 3).

The Riley Formation, central Texas (Palmer 1954).

The Honey Creek Formation, Wichita Mountains, Oklahoma (Frederickson 1946).

Greenland

The Bastion and Buen formations contain a total of three bradoriid species (Poulsen 1932; Siveter *et al.* 1996).

PALAEOGEOGRAPHICAL SETTING AND PALAEOECOLOGY

Palaeogeographical setting

During the Cambrian, much of present day North America formed the palaeocontinent of Laurentia which straddled the equator (Scotese and McKerrow 1990; McKerrow *et al.* 1992). In the early Cambrian, sedimentation along the margins of Laurentia was predominantly in relatively deep marine basins. Overstep onto the craton hardly occurred until the Mid Cambrian, which introduced the deposition of incomplete but fossiliferous clastic and carbonate sequences (Lochman-Balk 1971; Brasier and Cowie 1989).

Much of New Brunswick, Nova Scotia and eastern Newfoundland (western Newfoundland was part of Laurentia), together with southern Britain and parts of the north-eastern USA, formed components of palaeocontinental Avalonia (Scotese and McKerrow 1990; McKerrow *et al.* 1992), a microcontinent formed largely from late Precambrian island arcs. Avalonia was situated in the

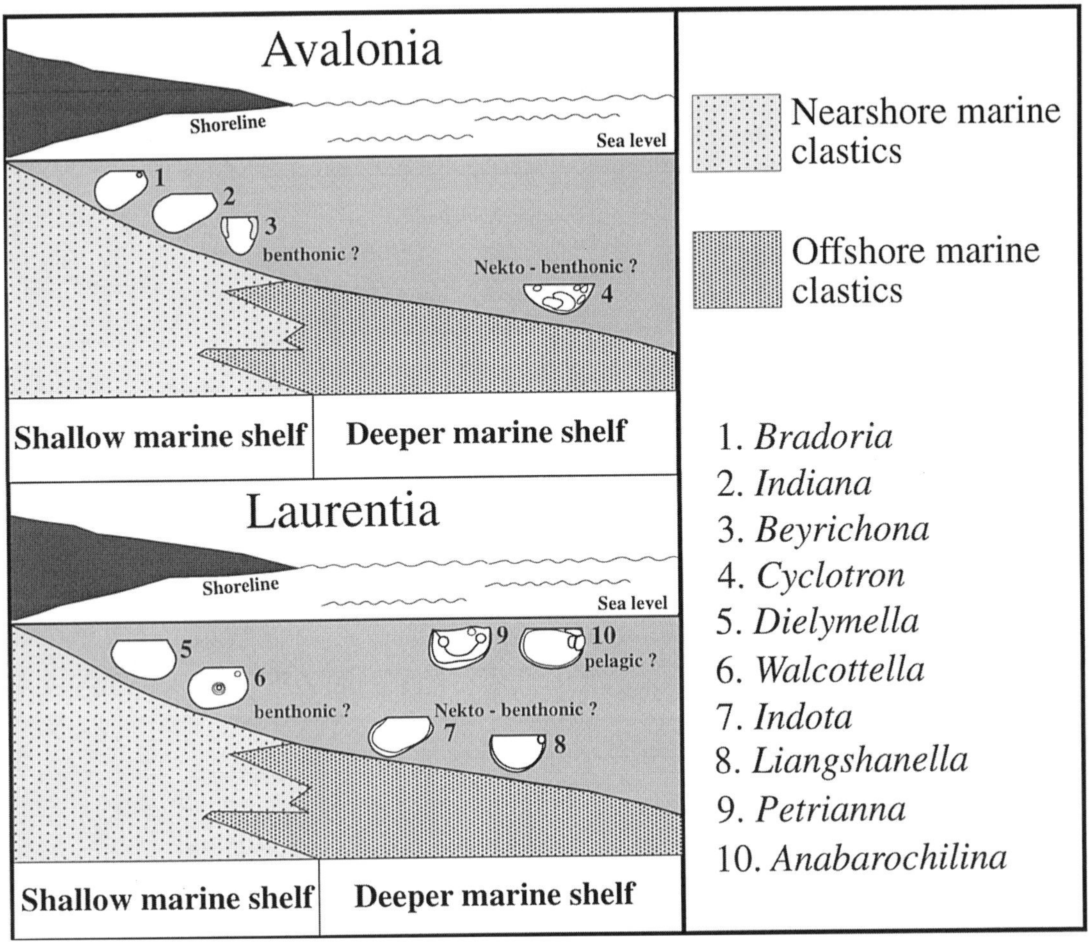

TEXT-FIG. 5. Possible habitat and mode of life of some North American and Greenland (*Petrianna* only) bradoriid and phosphatocopid genera occurring on the Cambrian palaeocontinents of Avalonia and Laurentia.

southern hemisphere, at high, temperate latitudes and was part of, or immediately adjacent to, Gondwana to the south (see Pickering and Smith 1995). The Early Cambrian sediments of Avalonia were deposited mainly in shallow marine environments; those of the later Cambrian accumulated in deeper-marine basins.

Palaeoecology of North American bradoriids and phosphatocopids

Like typical post-Cambrian Palaeozoic ostracods (e.g. see Siveter 1984; Williams and Siveter 1996), Cambrian bradoriid and phosphatocopid faunas appear to have occupied both shallow and deeper shelf marine settings (Text-fig. 5). Also like post-Cambrian ostracods, their patterns of geographical and facies distribution suggest that they may have been mostly benthic or nekto-benthic; contenders for a pelagic life-style are rare (e.g. see Hinz-Schallreuter 1993*b*; Siveter *et al.* 1996).

In the Avalonian sequences of eastern Canada bradoriid and phosphatocopid faunas occur in the marine shelf mudstones and sandstones of the Early Cambrian Hanford Brook Formation, the marine tuffaceous sandstones, siltstones and micaceous mudstones of the Mid Cambrian Dugald Formation and marine shelf sediments of the Maclean Brook Formation (Hutchinson 1952; Tanoli and Pickerill 1988; Landing and Westrop 1996). These shallow marine settings are characterized by species of *Bradoria*, *Beyrichona*, *Hipponicharion* and *Indiana* (Text-fig. 5). Faunas are typically of abundant specimens, but diversity is low (three to six species). Late Cambrian bradoriid and phosphatocopid faunas of Avalonia are characterized by phosphatocopids (e.g. *Cyclotron*) which seem to have adapted to deeper marine shelf settings (Taylor and Rushton 1971; Landing 1980; Martin and Dean 1988) and were possibly tolerant of oxygen-poor bottom conditions (Müller 1979; Williams *et al.* 1994*c*).

In the deeper marine shelf settings along the cratonic margin of Laurentia, as exemplified by the Early Cambrian Buen Formation of North Greenland (Siveter *et al.* 1996) and the Mid Cambrian Burgess Shale of British Columbia (Conway Morris 1986), faunas are generally of low diversity and are characterized by species of the cosmopolitan genera *Indota* Öpik, 1968, *Liangshanella* Huo, 1956 and cambriids (Text-fig. 5). Some elements of these faunas may have been pelagic, such as the svealutid *Anabarochilina* Abushik, 1960 and the cambriids (see Siveter *et al.* 1996). Within the strictly cratonic sequences of Laurentia, more shallow marine subtidal settings, as represented by the Mid Cambrian Bright Angel Shale of northern Arizona (Ulrich and Bassler 1931*a*; Lochman-Balk 1971), were characterized by low diversity (one or two species) presumed benthic assemblages of *Walcottella* and *Dielymella* (Text-fig. 5).

BIOSTRATIGRAPHY AND ZOOGEOGRAPHY

Biostratigraphy

Bradoriids and phosphatocopids occur throughout the Cambrian of North America. Most species are apparently short-ranging, often sparsely represented and currently are known to have only local geographical distribution (Text-figs 1–2, 6–7). However, a few species are known from coeval horizons, in Canada, southern Britain and Scandinavia and therefore have wider biostratigraphical value (e.g. Williams *et al.* 1994*c*; Rushton *et al.* in press).

Avalonia (Text-figs 2, 6). The stratotype for the Precambrian-Cambrian boundary is placed at the base of the *Phycodes pedum* trace fossil Biozone in the Avalon-Burin Peninsula of eastern Newfoundland (Brasier *et al.* 1994). Bradoriids first appear in the late Early Cambrian Hanford Brook Formation of New Brunswick and include the short-ranging *Beyrichona papilio*, *Beyrichona tinea* and *Hipponicharion eos* (all Matthew, 1886), and *Indiana secunda* (Matthew, 1895), which characterize strata of *Protolenus* trilobite Biozone age. Although useful for local correlation within New Brunswick (see Matthew 1898), these fossils are unrecorded from the coeval upper part of the Brigus Formation of Nova Scotia and eastern Newfoundland.

Several, apparently short-ranging species characterize Mid Cambrian rocks of Nova Scotia, including *Bradoria scrutator* Matthew, 1899 and *Indiana lippa* Matthew, 1902 in the Dugald Formation (*Harttella bucculenta* Biozone; see Landing 1996), and *Beyrichona triceps* Matthew, 1903 in the Maclean Brook Formation (*Paradoxides forchhammeri* trilobite Biozone). Their correlative potential in eastern Canada is limited because strata of equivalent age in New Brunswick and eastern Newfoundland are either absent, or lack or have poorly preserved bradoriid faunas. *B. triceps*, though apparently restricted to the Middle Cambrian in Nova Scotia, also occurs in the Tremadoc of Britain (Rushton *et al.* in press).

The late Cambrian *Cyclotron lapworthi* (Groom, 1902) occurs in the Maclean Brook Formation of eastern Newfoundland and also the *Olenus* trilobite Biozone in Britain and probably Scandinavia (Williams *et al.* 1994*c*; for revised stratigraphy see Landing 1996). '*Falites*' *pateli* Landing, 1980 is currently known only from the Late Cambrian part (*Parabolina spinulosa* trilobite Biozone) of the Chesley Drive Group of New Brunswick (Landing 1980, 1996).

The absence of bradoriids and phosphatocopids from the Avalonian Cambrian sequences of Rhode Island and Massachusetts (see Conway Morris 1989), may simply reflect a lack of sampling. This possibility is highlighted because several of the bradoriid and phosphatocopid species which occur in the Canadian Atlantic Provinces are also found at similar horizons elsewhere in southern

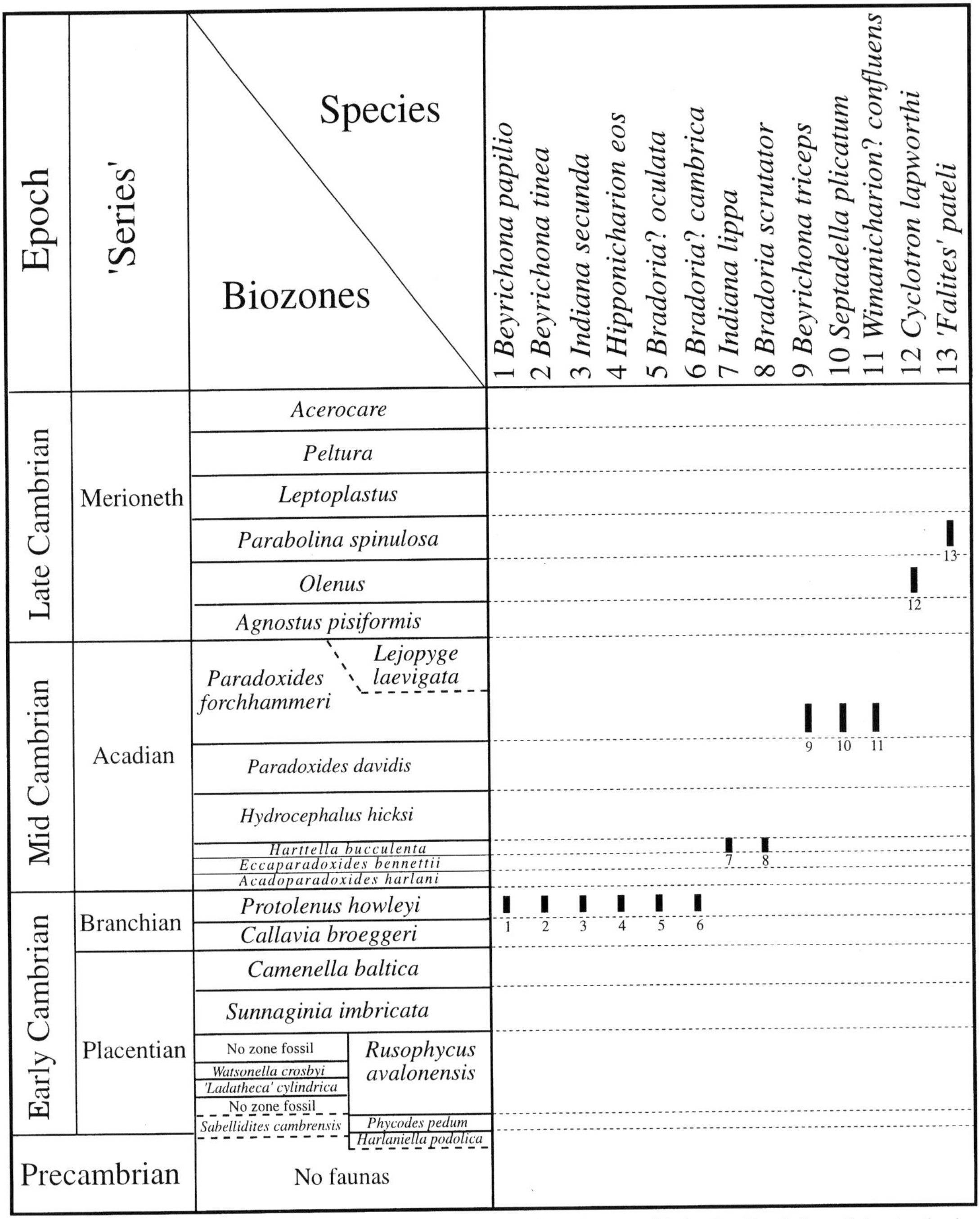

TEXT-FIG. 6. Stratigraphical distribution of bradoriids and phosphatocopids in the Cambrian of the Avalonian part (Canadian Atlantic Provinces) of North America.

Britain and Scandinavia. *Bradoria scrutator* occurs in the Mid Cambrian *Acrothele prima* Shale in England (Rushton *et al.* in press); *Cyclotron lapworthi* occurs in the Late Cambrian *Olenus* trilobite Biozone of England and probably Scandinavia (Williams *et al.* 1994*c*); and the Early Cambrian

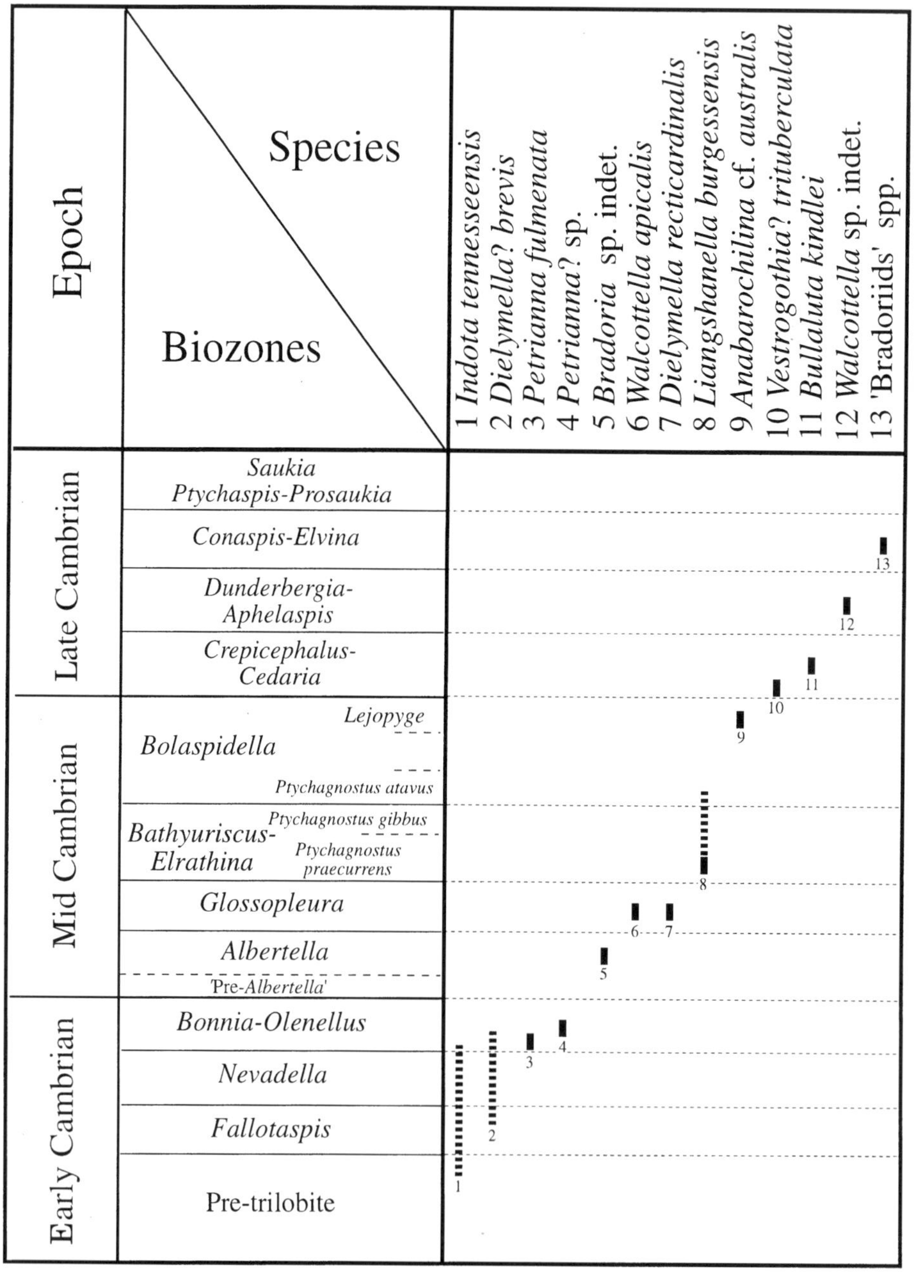

TEXT-FIG. 7. Stratigraphical distribution of bradoriids and phosphatocopids in the Cambrian of the Laurentian part of North America. The ranges of *Petrianna fulmenata* and *Petrianna*? sp. are based on records from Greenland (Siveter *et al.* 1996). Broken lines indicate uncertain occurrence.

(*Protolenus* trilobite Biozone); *Beyrichona tinea* is possibly conspecific with the ?Early Cambrian *Beyrichona gevalensis* Wiman, 1905 from Scandinavia.

Laurentia (Text-fig. 7). Most Laurentian bradoriid and phosphatocopid species are short-ranging and therefore offer correlative potential (Text-fig. 7). However, all of the species are known only from their type and nearby localities. The earliest well-documented and stratigraphically constrained species are those of the Early Cambrian (*Nevadella* to *Bonnia-Olenellus* trilobite biozones) Buen Formation of North Greenland, which include the short-ranging cambriids *Petrianna fulmenata* and *Petrianna*? sp. (Siveter *et al.* 1996). The Tennessee bradoriid *Indota tennesseensis* (Resser, 1938) appears to occur below the oldest olenellid trilobites (see Palmer 1971) and indicates an Early Cambrian age for the Chilhowee Group Murray Shale (Laurence and Palmer 1963). Undescribed early Cambrian bradoriid and phosphatocopid faunas of the Puerto Blanco Formation of Sonora, Mexico, include hipponicharionids (McMenamin and McMenamin 1990, fig. 4.29).

The Early Cambrian sequences of Vermont and New York State were probably deposited in a marine basin situated on the eastern margin of Laurentia, but their exact palaeotectonic setting is complicated by post-Cambrian tectonic events and is poorly constrained (see review in Pickering and Smith 1995). The Early Cambrian *Elliptocephala asaphoides* fauna in New York State has a few bradoriids, including *Anabarochilina rotundata* (Walcott, 1887), *Indiana dermatoides* (Walcott, 1887) and *Matthoria troyensis* (Ford, 1873) (see also Ulrich and Bassler 1931*a*; Lochman 1956). Although these species appear to be short-ranging and are not recorded from the younger *Acimetopus bilobatus* and *Pagetides* faunas (Lochman 1956; Palmer 1971), they presently have only local occurrence and biostratigraphical use.

Mid Cambrian species known only locally include: *Bradoria* sp. from shales of *Albertella* trilobite Biozone age in south-western Alberta (Bolton and Copeland 1963); *Dielymella recticardinalis* Ulrich and Bassler, 1931*a* and *Walcottella apicalis* Ulrich and Bassler, 1931*a* from the Bright Angel Shale of Arizona (*Glossopleura* trilobite Biozone); and *Anabarochilina* cf. *australis* from the Marjum and Weeks formations of Utah (*Lejopyge* trilobite Subzone, *Bolaspidella* trilobite Biozone: Hintze and Robison 1975; Robison 1989). *Liangshanella burgessensis* sp. nov., from the Mid Cambrian Burgess Shale of British Columbia (*Ptychagnostus praecurrens* trilobite Subzone: Robison 1984; Conway Morris 1986), may be long-ranging and therefore conspecific with specimens from the beds adjacent to the boundary between the Wheeler Formation and overlying Pierson Cove Formation of Utah (*Ptychagnostus atavus* trilobite Subzone, *Bolaspidella* trilobite Biozone: see Hintze and Robison 1975; Robison 1984; Text-fig. 7).

Amongst the rare late Cambrian bradoriid and phosphatocopid faunas are those of *Vestrogothia*? *trituberculata* (Lochman and Hu, 1960) from the Du Noir Limestone of Wyoming (*Cedaria* trilobite Biozone), *Walcottella* sp. from the Riley Formation of Texas (*Aphelaspis* trilobite Biozone; Palmer 1954), '*Leperditia*' *harrisi* Frederickson, 1946 from the Honey Creek Formation of Oklahoma (*Elvinia* trilobite Biozone; see also Lochman-Balk 1971) and *Bullaluta kindlei* Copeland, 1986 from the *Cedaria-Crepicephalus* Zone 5 conglomerate, western Newfoundland (see also Kindle 1981).

Zoogeography

During the Cambrian, Avalonia, Laurentia and Baltica were part of the Olenellid Trilobite Realm. The distribution of the bradoriid and phosphatocopid faunas of Avalonia and Laurentia broadly supports reconstructions of Cambrian palaeogeography, in which these two plates are shown separated by the Iapetus Ocean (e.g. Scotese and McKerrow 1990; McKerrow *et al.* 1992). Some taxa are provincial: for example, in the southern hemisphere, temperate latitude Avalonia and Baltica palaeocontinents the dominant bradoriid and phosphatocopid elements are genera such as *Beyrichona*, *Cyclotron*, *Falites* Müller, 1964, *Hipponicharion*, *Indiana* and *Wimanicharion*. The shallow marine, cratonic bradoriid and phosphatocopid faunas of low latitude, tropical Laurentia also appear to be strongly provincial, being characterized by *Dielymella* and *Walcottella*.

Some bradoriids and phosphatocopids occur widely in the Olenellid Realm; for example, *Bradoria* in the lower Middle Cambrian of Alberta (Laurentia), Nova Scotia and southern Britain (both Avalonia). Moreover, those bradoriid and phosphatocopid faunas occupying relatively deeper marine shelf settings on Laurentia, along the margins of the craton, include at least some cosmopolitan genera. For example, *Indota* is known also from Australia (Öpik 1968) and *Liangshanella* from China (Huo *et al.* 1991); *Anabarochilina* occurs world-wide in the upper Middle and lower Upper Cambrian (Siveter *et al.* 1993; Melnikova *et al.* in press); furthermore, *Petrianna* (Siveter *et al.* 1996), from Greenland, bears close affinity to cambriid bradoriids from Siberia and south-west China.

Although poorly known at generic level, the Early Cambrian bradoriid faunas of New York State appear to show affinities with contemporaneous Avalonian faunas (*Indiana*, *Matthoria*) rather than with Laurentian faunas (see also Palmer 1971).

SYSTEMATIC PALAEONTOLOGY

Phylum ARTHROPODA Siebold and Stannius, 1845
Order BRADORIIDA Raymond, 1935 [*nom corr.* Ivanova 1960]

Remarks. Raymond (1935) established the order Bradoriida for Cambrian ostracod-like fossils, and conventionally all of these bivalved arthropods have been considered to be ostracods (e.g. Sylvester-Bradley 1961; Müller 1964, 1979; Jones and McKenzie 1980). Müller (1964) subdivided the Bradoriida into the suborders Phosphatocopina, which he considered to have primary phosphatic carapaces, and Bradoriina, having supposed chitinous or calcareous carapaces. Both taxa were later raised to ordinal level (Müller 1982), a scheme adopted herein.

Jones and McKenzie (1980) considered Bradoriida *s.l.* to be probably a polyphyletic taxon, embracing various arthropod groups including possibly some ancestral ostracods. Based on evidence from preserved soft parts, phosphatocopids are now thought to represent advanced stem-group Crustacea (e.g. Müller and Walossek 1991; Walossek and Müller 1992). Bradoriids with preserved soft parts are very rare but indicate that at least some members of the group are not ostracods, or similar to phosphatocopids (Hou *et al.* 1996).

In the absence of soft part preservation, bradoriid and phosphatocopid taxonomy is necessarily based on gross carapace morphology. Carapace composition appears to be an unreliable criterion for distinguishing bradoriids from phosphatocopids (e.g. Müller 1964, 1979; cf. Hinz-Schallreuter 1993*b*; Siveter and Williams 1995; Siveter *et al.* 1996 and herein); many Bradoriida are now known to have a phosphatic carapace (e.g. *Beyrichona*, *Hipponicharion*, *Indiana*, *Liangshanella*, *Petrianna*). Bradoriid carapaces are generally relatively large (adults *c.* 2·5–17·5 mm long), postplete, have ornament of granulostriation, reticulation or punctation, lack an interdorsum, and rarely have a mineralized inner lamella. Phosphatocopids are typically small (adults *c.* 1–5 mm long), preplete, smooth, and often possess an interdorsum and well preserved inner lamella. Lobal morphology, which varies widely both in phosphatocopids and bradoriids, is used as a basis for familial and subfamilial classification within both groups (e.g. Hinz-Schallreuter 1993*b*; Siveter *et al.* 1996).

Family BRADORIIDAE Matthew, 1902

1902 Bradoriidae; Matthew, p. 443 [*partim*], *non Beyrichona* Matthew, 1886 [= Beyrichonidae Ulrich and Bassler, 1931*a*].
1903 Bradoriidae; Matthew, p. 155 [*partim*], *non Beyrichona* Matthew, 1886 [= Beyrichonidae Ulrich and Bassler, 1931*a*].
1931*a* Bradoriidae Matthew; Ulrich and Bassler, p. 12.
1931*a* Indianidae new family; Ulrich and Bassler, p. 68 [*partim*], *non Dielymella* Ulrich and Bassler, 1931*a*.
1931*b* Indianitidae new name; Ulrich and Bassler, p. 364.
1961 Bradoriidae Matthew, 1902; Sylvester-Bradley, p. Q102.
1961 Indianidae Ulrich and Bassler, 1931; Sylvester-Bradley, p. Q103 [*partim*], ?*Mononotella* Ulrich and Bassler 1931*a*; *non Dielymella* Ulrich and Bassler, 1931*a*.
1961 Bradoriidae Matthew; Öpik, p. 176.
1961 Indianidae Ulrich and Bassler; Öpik, p. 176.
1968 Indianidae Ulrich and Bassler; Öpik, p. 17.
1968 Indianitidae; Öpik, p. 17.

1968 Bradoriidae Matthew, 1903 [*sic*]; Öpik, p. 21.
1974 Bradoriidae Matthew, 1902; Kozur, p. 826 [*partim*], *non Falites* Müller, 1964 [= Falitidae Müller, 1964].
1974 Indianidae Ulrich and Bassler, 1931*a*; Kozur, p. 826 [*partim*], ?*Mononotella* Ulrich and Bassler 1931*a*; *non Vestrogothia granulata* Müller, 1964 [= phosphatocopid], *Dielymella* Ulrich and Bassler, 1931*a*.
1981 Bradoriidae Matthew (1902]); Gründel, p. 60, *non Waldoria* gen. nov. [= phosphatocopid], *Falites* Müller, 1964 [= Falitidae Müller, 1964].
1993*b* Bradoriidae Matthew, 1902; Hinz-Schallreuter, p. 413.

Diagnosis. Medium sized (adults *c*. 6–7 mm long), postplete Bradoriida with thin shelled, subequivalved carapace without posterior gape. Dorsum straight; hinge-line well developed, shorter than carapace length. Commissure straight. Node(s) well developed mid-anteriorly or anterodorsally or absent. When present, marginal structure is a very narrow ridge. Valves have concentric striation/reticulostriation, punctation or are smooth.

Remarks. Bradoriidae, such as *Bradoria* Matthew, 1899 and *Walcottella* Ulrich and Bassler, 1931*a*, have traditionally been distinguished from Indianidae Ulrich and Bassler, 1931*a* by having an anterodorsal node and supposedly less posterior and anterior gape to the valves (Ulrich and Bassler 1931*a*, pp. 68–69; see also Öpik 1968). Our investigations of typical indianids (e.g. type species *Indiana lippa* Matthew, 1902; *Indiana lentiformis* (Cobbold, 1921); see Siveter and Williams 1995) indicate no evidence for a posterior gape to the carapace. Furthermore, the presence of an anterodorsal node varies between congeneric species of Bradoriidae (e.g. in *Bradoria*). Typical species of *Indiana* tend to be more elongately postplete than those of *Bradoria* (cf. Pls 1, 3), but we consider that this shape difference alone is insufficient to sustain a distinction between Bradoriidae and Indianidae, and therefore regard these families as synonymous. Because of its preplete outline we exclude *Dielymella* Ulrich and Bassler, 1931*a* from Bradoriidae (cf. Ulrich and Bassler 1931*a*, pp. 85–86).

Hinz-Schallreuter (1993*b*) referred her monospecific genus, *Ifersiktia* from the Middle Cambrian of Morocco, to the Bradoriidae. The holotype (and only specimen) of the type species, *I. mipora*, is a single incomplete carapace which differs from typical Bradoriidae by having a weak anterodorsal sulcus. The systematic position of the Chinese genera which have been referred to Bradoriidae and Indianidae (e.g. see Shu 1990*b*, with nomenclatorial annotations by Malz 1990) needs investigation.

Genus BRADORIA Matthew, 1899

1899 *Bradoria* n. gen., Matthew, p. 204.
1902 *Bradorona*, n. subgen., Matthew, p. 444.
1902 *Bradoria*; Matthew, p. 451.
1903 *Bradorona*; Matthew, p. 156.
1903 *Bradoria*; Matthew, p. 162.
1931*a* *Bradoria* Matthew; Ulrich and Bassler, p. 12.
1945 *Bradoria* Matthew, 1899; Resser, p. 216.
1961 *Bradoria* Matthew, 1899; Sylvester-Bradley, p. Q102.
1961 *Bradorona* Matthew, 1903 [*sic*]; Sylvester-Bradley, p. Q102.
1968 *Bradoria* Matthew, 1899; Öpik, p. 22 [see text below].
1993*b* *Bradoria* Matthew, 1899; Hinz-Schallreuter, p. 413.
1993*b* *Bradorona* Matthew, 1902; Hinz-Schallreuter, p. 413.

Type species. Subsequently designated Ulrich and Bassler (1931*a*, p. 15); *Bradoria scrutator* Matthew, 1899, from the Middle Cambrian of Nova Scotia.

Diagnosis. Bradoriids with a posterodorsal curve and a more pronounced anterior curve in the lateral outline of the valve. Lateral outline is fairly straight between these curves and the cardinal corners, and ventrally convex to strongly convex. Sub-circular anterodorsal node. Valves typically have concentric striation/reticulostriation to punctation. Very narrow ridge may be present immediately adjacent to free margin. Adults *c.* 6 mm long.

Remarks. Among specimens from the Canadian Atlantic provinces that have been compared to *Bradoria*, those cited by Ulrich and Bassler (1931*a*, pp. 29–30, 100) under *Leperditia steadi* Matthew, 1890 (ROM 818CM) and *Leperditia ventricosa* Matthew, 1890 (ROM 804CM) are probably trilobite glabellae; *Beyrichia*(?) *primaeva* Matthew, 1898 (ROM 127CM) is probably the external mould of the dorsal portion of a specimen of *Beyrichona tinea* Matthew, 1886; and the single specimen on which *Bradoria*(?) *ornata* Matthew, 1902 is based is an indeterminate steinkern (ROM 73CM). *Bradoria*? *curta* Ulrich and Bassler, 1931*a* is based on a single, anterodorsally damaged, indeterminate specimen (USNM 50547). Specimen(s) on which *Primitia aurora* Matthew, 1894 were based (referred to *Bradoria*? by Matthew 1899; see also Ulrich and Bassler 1931*a*) have not been located and no further material is known. The punctate *Bradoria tontoensis* Ulrich and Bassler, 1931*a* (p. 24, pl. 3, fig. 20), based on a single incomplete and poorly preserved valve from the Mid Cambrian Bright Angel Shale, Arizona (USNM 81377), might be a fragment of *Walcottella* (see also Resser 1945, p. 217, pl. 27, fig. 4).

More than 50 species and subspecies have been referred to *Bradoria* (see Kempf 1986). Those from outside the Avalonian areas of eastern Canada and southern Britain need reassessment. For example, *Bradoria mapora* Hinz-Schallreuter, 1993*b* from the Middle Cambrian of Morocco is based on a single incomplete valve. Species referred to *Bradoria* from the former Soviet Union (e.g. *Bradoria*? *estonica* Melnikova, 1987, *Bradoria sayutinae* Melnikova, 1988) have well-developed marginal ridges and therefore appear to belong outside *Bradoria* (Melnikova *et al.* in press). Of the four co-occurring (conspecific?) species of '*Bradoria*' from the Lower Cambrian of Australia (Öpik 1968, pp. 22–26, pls 1–2), most consist of flattened, strongly deformed specimens; we consider that only *Bradoria cornulata* Öpik, 1968, with its sub-circular anterodorsal node and a similar lateral outline to *Bradoria scrutator*, might be congeneric with *Bradoria*. *Bradoria* sp. of Bolton and Copeland (1963), from the Middle Cambrian of western Canada, lacks an anterodorsal node. The affinities of *Bradoria* sp. of Andres (1969, p. 76, fig. 12), based on a small (juvenile?), smooth, damaged left valve from the Middle Cambrian of Scandinavia, are uncertain.

Occurrence. Lower Cambrian of New Brunswick? (Ulrich and Bassler 1931*a*) and lower Middle Cambrian of Nova Scotia (Matthew 1899 and herein), Alberta (Copeland and Bolton 1963) and southern Britain (Cobbold 1934). All other occurrences (e.g. Öpik 1968; Hinz-Schallreuter 1993*b*) need verification.

EXPLANATION OF PLATE 1

Figs 1–14. *Bradoria scrutator* Matthew, 1899; Mid Cambrian Dugald Formation, Dugald Brook (figs 1, 3–5, 7, 9–14) and Boundary Brook (figs 2, 6, 8), Cape Breton, Nova Scotia; all lateral views. 1, USNM 487774; partly exfoliated left valve (lectotype of *Bradoria benepuncta spissa* Ulrich and Bassler, 1931*a*, pl. 2, fig. 8), stereo-pair; ×8. 2, ROM 93CM(A); anterodorsally and posterodorsally incomplete, partly exfoliated left valve (lectotype of *Bradorona perspicator* Matthew, 1902, pl. 1, fig. 8*a–d*); ×9. 3, USNM 486353; partly exfoliated left valve; ×10. 4, USNM 486355; left valve of partly exfoliated carapace; ×10. 5, ROM 144CM(A); incomplete, partly exfoliated right valve (lectotype of *Bradoria vigilans* Matthew, 1899, pl. 4, fig. 2*a–c*); ×10. 6, ROM 105CM D; anterodorsally exfoliated right valve exposing possible anterodorsal muscle attachment scar (lectotype of *Bradorona spectator* Matthew, 1902); ×9. 7, ROM 78CM; partly exfoliated, incomplete carapace (holotype of *Bradorona spectator spinosa* Matthew, 1902, pl. 1, fig. 13*a–b*); ×9. 8, ROM 71CM A; partly exfoliated, posteriorly incomplete right valve (lectotype of *Bradoria observator benepuncta* Matthew, 1902); ×9. 9, 11, ROM 142CM C; partly exfoliated, ventrally and posteriorly incomplete left valve, lectotype (Matthew 1899, pl. 4, fig. 1*a–c*). 9, stereo-pair; ×10. 11, close-up of puncta; ×17. 10, ROM 142CM B; partly exfoliated right valve (paralectotype of *Bradoria scrutator* Matthew, 1899), stereo-pair; ×10. 12, USNM 486352; incomplete, open carapace (Ulrich and Bassler 1931*a*, pl. 2, fig. 3); ×8. 13, ROM 689CM; left valve of ventrally incomplete carapace (holotype of *Bradoria vigilans obesa* Matthew, 1902); ×10. 14, USNM 483189; anterodorsally incomplete right valve (Ulrich and Bassler 1931*a*, pl. 2, figs 1–2), stereo-pair; ×8.

SIVETER and WILLIAMS, *Bradoria*

Bradoria scrutator Matthew, 1899

Plate 1, figures 1–14; Text-fig. 8

1899 *Bradoria scrutator*, n. sp., Matthew, p. 204, pl. 4, fig. 1*a–c*.
1899 *Bradoria vigilans*, n. sp., Matthew, p. 205, pl. 4, fig. 2*a–c*.
?1899 *Bradoria rugulosa*, n. sp., Matthew, p. 205, pl. 3, fig. 3*a–d*.
1899 *Schmidtella acuta*, n. sp., Matthew, p. 206, pl. 4, fig. 3*a–c*.
?1902 *Leperditia*?? *rugosa*, n. sp., Matthew, pl. 1, fig. 7*a–c*.
1902 *Bradorona perspicator*, n. sp., Matthew, p. 444, pl. 1, fig. 8*a–d*.
1902 *Bradorona perspicator* mutation *maxima*, n. mut., Matthew, p. 445, pl. 1, fig. 9*a–b*.
?1902 *Bradorona perspicator* mutation *magna*, n. mut., Matthew, p. 446, pl. 1, fig. 11*a–b*.
?1902 *Bradorona perspicator* mutation *major*, n. mut., Matthew, p. 446, pl. 1, fig. 10*a–b*.
1902 *Bradorona spectator*, n. sp., Matthew, p. 447, pl. 1, fig. 12*a–d*.
?1902 *Bradorona spectator* var. *acuta*, n. var., Matthew, p. 447.
1902 *Bradorona spectator* mutation *spinosa*, n. mut., Matthew, p. 448, pl. 1, fig. 13*a–b*.
?1902 *Bradorona spectator* mutation *aequata*, n. mut., Matthew, p. 448, pl. 1, fig. 14*a–b*.
1902 *Bradorona observator*, n. sp., Matthew, p. 448, pl. 1, fig. 15a–c.
1902 *Bradorona observata* var. *benepuncta*, n. var., Matthew, p. 449, pl. 1, fig. 16.
1902 *Bradorona observator* mutation *laevis*, n. mut., Matthew, p. 450.
?1902 *Bradorona observator* mutation *ligata*, n. mut., Matthew, p. 451, pl. 1, fig. 17 [specimens not traced].
1902 *Bradoria scrutator*; Matthew, p. 452, pl. 2, fig. 1*a–c*.
1902 *Bradoria vigilans*; Matthew, p. 454, pl. 2, fig. 2*a–c*.
1902 *Bradoria vigilans* mut. *obesa*, n. mut., Matthew, p. 455.
?1902 *Bradoria rugulosa*; Matthew, p. 456, pl. 2, fig. 3*a–d*.
?1902 *Indiana ovalis*, n. sp., Matthew, p. 461, pl. 2, fig. 8*a–c*.
1902 *Schmidtella*(?) *pervetus* mutation *concinna*, n. mut., Matthew, p. 464.
1902 *Schmidtella acuta*; Matthew, p. 465, pl. 2, fig. 12*a–c*.
1903 *Bradorona perspicator*; Matthew, p. 156, pl. 7, fig. 8*a–d*.
1903 *Bradorona perspicator* mutation *maxima*; Matthew, p. 157, pl. 7, fig. 9*a–b*.
?1903 *Bradorona perspicator* mutation *magna*; Matthew, p. 158, pl. 7, fig. 11*a–b*.
?1903 *Bradorona perspicator* mutation *major*; Matthew, p. 158, pl. 7, fig. 10*a–b*.
1903 *Bradorona spectator*; Matthew, p. 158, pl. 7, fig. 12*a–d*.
?1903 *Bradorona spectator* variety *acuta*; Matthew, p. 159.
1903 *Bradorona spectator* mutation *spinosa*; Matthew, p. 159, pl. 7, fig. 13*a–b*.
?1903 *Bradorona spectator* mutation *aequata*; Matthew, p. 160, pl. 7, fig. 14*a–b*.
1903 *Bradorona observator*; Matthew, p. 160, pl. 7, fig. 15*a–c*.
1903 *Bradorona observata* variety *benepuncta*; Matthew, p. 161, pl. 7, fig. 16.
1903 *Bradorona observator* mutation *laevis*; Matthew, p. 161.
?1903 *Bradorona observator* mutation *ligata*; Matthew, p. 162, pl. 7, fig. 17.
1903 *Bradoria scrutator*; Matthew, p. 163, pl. 8, fig. 1*a–c*.
1903 *Bradoria vigilans*; Matthew, p. 164, pl. 8, fig. 2*a–c*.
1903 *Bradoria vigilans* mut. *obesa*; Matthew, p. 165.
?1903 *Bradoria rugulosa*; Matthew, p. 166, pl. 8, fig. 3*a–d*.
?1903 *Indiana ovalis*; Matthew, p. 170, pl. 8, fig. 8*a–c*.
1903 *Schmidtella*(?) *pervetus* mutation *concinna*, n. mut., Matthew, p. 173.
1903 *Schmidtella*(?) *acuta*; Matthew, p. 173, pl. 8, fig. 12*a–c*.
1905 *Brodorona spectatrix* mut. *spinosa* Matt [*sic*]; Wiman, p. 47.
1931*a* *Bradoria scrutator* Matthew; Ulrich and Bassler, p. 14, pl. 1, fig. 11, ?fig. 12; pl. 3, fig. 11.
1931*a* *Bradoria vigilans* Matthew; Ulrich and Bassler, p. 16, pl. 1, fig. 15.
1931*a* *Bradoria vigilans concentrica*, new variety, Ulrich and Bassler, p. 17, pl. 3, figs 18–19.
1931*a* *Bradoria benepuncta* (Matthew); Ulrich and Bassler, p. 18, pl. 1, fig. 16; pl. 2, figs 6–7; pl. 3, fig. 7.
1931*a* *Bradoria benepuncta spissa*, new variety, Ulrich and Bassler, p. 18, pl. 2, figs 8–9.
1931*a* *Bradoria medialis*, new species, Ulrich and Bassler, p. 19, pl. 3, fig. 5.
1931*a* *Bradoria acuta* (Matthew); Ulrich and Bassler, p. 19, pl. 1, fig. 25; pl. 3, figs 1–2.
?1931*a* *Bradoria subquadrata*, new name, Ulrich and Bassler, p. 20, pl. 3, fig. 3.

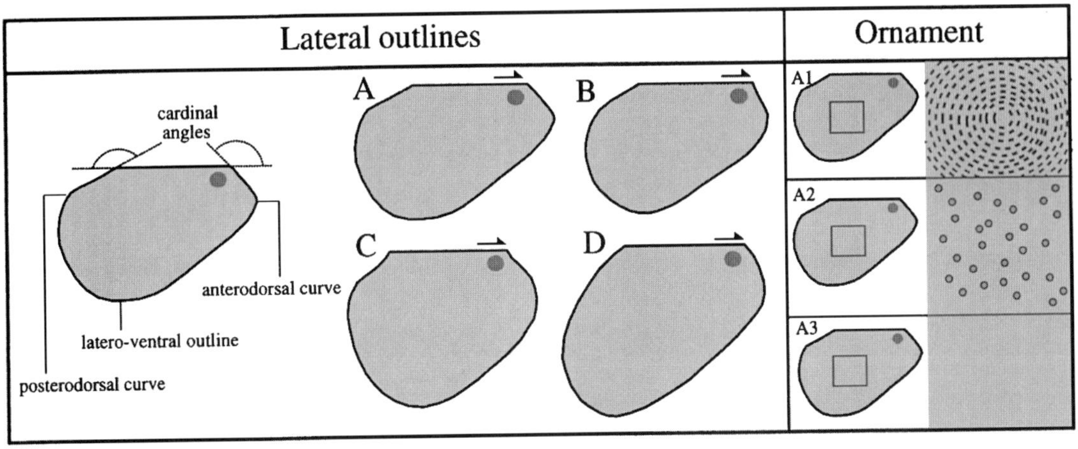

TEXT-FIG. 8. Variation in *Bradoria scrutator* Matthew, 1899. A–D are typical lateral outlines between which variation is continuous. Variation of ornament is due to preservational factors (more abraded valves are punctate or smooth): Al, concentric striation; A2, punctation; A3, smooth. Particular varieties do not cluster at certain horizons or localities.

?1931*a* *Bradoria* sp.; Ulrich and Bassler, p. 20.
?1931*a* *Bradoria elongata*, new species, Ulrich and Bassler, p. 20, pl. 3, fig. 8.
1931*a* *Bradoria ovalis* (Matthew); Ulrich and Bassler, p. 21 [*partim*; *non Indiana ovalis* Matthew, 1902; *non Indiana ovalis prima* Matthew, 1902 = *nomina dubia*], pl. 3, fig. 6; *non* pl. 1, fig. 18 [= *I. ovalis*].
?1931*a* *Bradoria rugulosa* Matthew; Ulrich and Bassler, p. 21, pl. 1, fig. 17.
1931*a* *Bradoria robusta* (Matthew); Ulrich and Bassler, p. 22 [*partim*; *non Aparchites*(?) *robustus* [Matthew, 1898] = *nomen dubium*], pl. 1, fig. 3; pl. 2, figs 1–4; ?pl. 1, figs 2, 4–5, *non* pl. 1, fig. 1 [= *Aparchites*(?) *robustus*].
1931*a* *Bradoria robusta longula*, new variety, Ulrich and Bassler, p. 24, pl. 2, fig. 5.
1931*a* *Bradoria perspicator* (Matthew) *s.s.*; Ulrich and Bassler, p. 24, pl. 1, figs 6–9; pl. 3, fig, 12.
1931*a* *Bradoria spectator* (Matthew) *s.s.*; Ulrich and Bassler, p. 25, pl. 1, fig. 19.
1931*a* *Bradoria spectator spinosa* (Matthew); Ulrich and Bassler, p. 26, pl. 1, fig. 20, ?fig. 21.
1931*a* *Bradoria obesa* (Matthew); Ulrich and Bassler, p. 27, pl. 1, fig. 27; pl. 3, fig. 13.
1931*a* *Bradoria concinna* (Matthew); Ulrich and Bassler, p. 28, pl. 3, fig. 9.
1934 *Bradoria benepunctata* (Matthew) [*sic*]; Cobbold, p. 379, pl. 41, fig. 7.
1956 *Bradoria benepuncta* (Matthew); Lochman, p. 1356.
1961 *B. scrutator*; Sylvester-Bradley, p. Q102, fig. 37.1*a*.
1961 *B. robusta*; Sylvester-Bradley, p. Q102, fig. 37.1*b*.
1993*b* *Bradoria scrutator* Matthew, 1899; Hinz-Schallreuter, p. 413, fig. 11.4.
1993*b* *Bradorona perspicator* Matthew, 1902; Hinz-Schallreuter, p. 413, fig. 13.1.
1993*b* *Bradorona* sp.; Hinz-Schallreuter, fig. 13.2.

Lectotype. Subsequently designated Ulrich and Bassler (1931*a*, p. 15, pl. 1, fig. 11); ROM 142CM C (Matthew 1899, pl. 4, fig. 1*a–c*; Pl. 1, figs 9, 11), a partially exfoliated juvenile left valve with incomplete ventral and posterior margins; early Mid Cambrian Dugald Formation (division E3e; see Text-fig. 3), Bourinot Group, Dugald Brook, Eskasonie, Cape Breton, Nova Scotia.

Material. Some 80 Canadian specimens; mostly incomplete and/or abraded valves. ROM 71CM A–B; 75CM A–G; 77CM A–C; 78CM; 88CM; 93CM(A); 93CM(B) B–D; 105CM A–G; 107CM A–C; 142CM A–C; 144CM(A); 144CM(B); 159CM A–B; 689CM; 26292 A–B; NBMG 3305, 4390/1, 4390/2, 9191/1, 9191/2, 9192, and 9898–9901; USNM 50518, 50527, 50532, 50535–50536, 50540–50542, 50544, 50548, 81379, 483181–483189, 483193–483194, 483883–483887, 483891, 486352–486356, 487774. Possibly conspecific are ROM 76CM A–B, 90CM, 92CM, 131CM, 551CM, 26291C; NBMG 9902, 9930–34; and USNM 483195–483196.

Measurements. Cape Breton valves range from 1·33 mm to 6·67 mm long. A British right valve (BGS 51691) is 3·24 mm long.

Diagnosis. Species of *Bradoria* with pronounced anterior curve of lateral valve outline, a sub-circular anterodorsal node and fine, concentric striation.

Description. Postplete carapace lacking posterior gape. Dorsal margin straight, shorter than valve length. Hinge-line well developed. Lateral outline has pronounced anterior curve and more weakly developed posterodorsal curve. Anterior cardinal angle generally greater than posterior cardinal angle. Ventral part of lateral outline evenly convex to weakly acuminate slightly posterior of mid-length. Sub-circular node near anterocardinal corner. Very narrow ridge present immediately adjacent to free margin. Almost entire lateral surface of valves have concentric striation.

Remarks. In lateral view *B. scrutator* shows marked variation in outline, especially affecting the anterior and posterior cardinal angles, anterior and posterodorsal curves and its length-height ratio (Text-fig. 8). However, particular variants do not cluster at specific horizons or localities. In well preserved *B. scrutator* ornament consists of concentric striation, but abraded valves are punctate or smooth (see Text-fig. 8).

Matthew (1899, 1902, 1903) perceived a progressive change in faunas from the Dugald Formation, from where he described some 25 species, varieties and 'mutations' of bradoriids which we herein consider to be junior synonyms or possible synonyms of *B. scrutator*. Indeed, subsequent authors (e.g. Hutchinson 1952 and references therein) have emphasized that the fauna is essentially similar throughout the formation. Many of the bradoriid taxa distinguished by Matthew (1899, 1902, 1903) and Ulrich and Bassler (1931*a*), on the basis of a diagnostic lateral valve outline and anterodorsal node, are incomplete specimens. Furthermore, some of these taxa, such as *Schmidtella*(?) *pervetus concinna* Matthew, 1902, are known only from a single, often incomplete or incompletely exposed valve.

Hinz-Schallreuter (1993*b*) preferred to limit *B. scrutator* to the lectotype. However, Matthew (1899) based *B. scrutator* on four specimens, three of which, all conspecific juveniles (< 3·00 mm long), we have identified (cf. Ulrich and Bassler 1931*a*, p. 15): a largely exfoliated right valve preserving much of the original lateral shape (ROM 142CM A); a partially exfoliated right valve with an incomplete free margin (ROM 142CM B; Pl. 1, fig. 10); and the lectotype (ROM 142CM C).

We regard the following taxa as synonyms of *B. scrutator*:

1. *Bradoria vigilans*, *B. rugulosa* and *Schmidtella acuta* (all Matthew, 1899), described originally from the same lithostratigraphical divison and locality as *B. scrutator* (Matthew 1899, 1902, 1903). *B. vigilans* was based on four incomplete juvenile specimens: two right valves, ROM 144CM(A) (also labelled E; Matthew 1899, pl. 4, fig. 2*a–c*; designated lectotype by Ulrich and Bassler 1931*a*; Pl. 1, fig. 5) and ROM 144CM(B) (also labelled D); and a left valve and a carapace, ROM 26292A–B. Syntypes of *B. rugulosa* are part and counterpart of a small juvenile valve (ROM 107CM A–B; possibly Matthew 1899, pl. 4, fig. 3*a–c*; herein designated lectotype) and a poorly preserved, incomplete juvenile carapace (ROM 107CM C). Syntypes of *S. acuta* are an almost complete right valve (ROM 159CM A; herein designated lectotype) and an anteriorly incomplete juvenile (ROM 159CM B).
2. *Bradoria vigilans obesa* Matthew, 1902 (holotype by monotypy = ROM 689CM; Pl. 1, fig. 13; ventrally incomplete juvenile carapace from divison E1b at Dugald Brook) and *Schmidtella*(?) *pervetus concinna* Matthew, 1902 (holotype by monotypy = probably specimen ROM 88CM; small juvenile left valve from division E1d at Boundary Brook). '*Bradoria concinna*' of Ulrich and Bassler 1931*a*, p. 28 (USNM 50510) is indeterminate.
3. *Bradorona perspicator* (type species), *B. spectator*, *B. spectator spinosa*, *B. observator*, *B. observator benepuncta* and *B. observator laevis* (all Matthew, 1902). Syntypes of *B. perspicator* comprise an antero- and posterodorsally incomplete left valve from divison E1d, Dugald Formation

at Boundary Brook (ROM 93CM(A); Matthew 1902, pl. 8, fig. 8*a–d*; Pl. 1, fig. 2; designated lectotype by Ulrich and Bassler 1931*a*, p. 25); a mostly complete right valve from division E1a, Dugald Formation at Boundary Brook (ROM 93CM(B) B); an external mould (ROM 93CM(B) C = counterpart of lectotype ROM 93CM(A)); and an indeterminate valve from division E1c at Dugald Brook (ROM 93CM(B) D). Identified syntypes of *B. spectator* consist of seven juveniles (ROM 105CM A–G), from Dugald Brook (division E3d) and Boundary Brook (division E1d; Matthew 1902, p. 447). Specimen ROM 105CM E is possibly the original of Matthew 1902, plate 1, figure 12*a–d*; syntype ROM 105CM D is herein designated lectotype (Pl. 1, fig. 6). *B. spectator spinosa*, from divison E1e at Dugald Brook, is based on an incomplete carapace (ROM 78CM, holotype by monotypy; left valve figured by Matthew 1902, pl. 1, fig. 13*a–b*; Pl. 1, fig. 7). Syntypes of *B. observator* consist of at least seven juveniles (ROM 75CM A–G), from division E1d at Boundary Brook; specimen ROM 75CM C is possibly the original of Matthew 1902, plate 1, figure 15*a–c* and is herein designated lectotype. A conspecific, possible additional syntype is the unfigured NBMG 3305, labelled *B. observator* and 'type'. Syntypes of *B. observator benepuncta*, from Boundary Brook (specimens labelled division 'Ela'; given as E1d in Matthew 1902), are a posteriorly incomplete right valve (ROM 71CM A; herein designated lectotype, Pl. 1, fig. 8) and a slightly damaged carapace (ROM 71CM B; Matthew 1902, pl. 1, fig. 16). Syntypes of *B. observator laevis* are an incomplete external mould (ROM 77CM A), an incomplete carapace (ROM 77CM B), and an incomplete right valve (ROM 77CM C; herein designated lectotype), from divisions E1b and E2c at Dugald Brook.

4. The following taxa erected by Ulrich and Bassler 1931*a*: *Bradoria subquadrata* (see below). *Bradoria vigilans concentrica* from division E3d at Dugald Brook; five syntypes (*ex* USNM 50527): incomplete right valve, retained as USNM 50527 (labelled 'holotype', but not identified in Ulrich and Bassler 1931*a*; herein designated lectotype); a dorsal fragment and an incomplete carapace (USNM 483185–483186; Ulrich and Bassler 1931*a*, pl. 3, figs 18–19); and unfigured left and right valves (USNM 483187–483188). *Bradoria benepuncta spissa* (syntypes: USNM 487774, USNM 50544; two mostly complete left valves, Ulrich and Bassler 1931*a*, pl. 2, figs 8–9; USNM 487774, designated lectotype herein; Pl. 1, fig. 1), *B. medialis* (USNM 50548, complete right valve; holotype by monotypy; Ulrich and Bassler 1931*a*, pl. 3, fig. 5) and *B. elongata* (USNM 50541, ventrally incomplete left valve; holotype by monotypy; Ulrich and Bassler 1931*a*, pl. 3, fig. 8), all from division E2b at Dugald Brook. *Bradoria robusta longula* (USNM 81379, incomplete left valve; holotype by monotypy) from division Elb at Dugald Brook.

The following taxa of Matthew (1902) may also be synonymous with *B. scrutator*, but preservational aspects preclude firm identification: *Bradorona perspicator maxima* (ROM 90CM; incomplete right valve; holotype by monotypy, Matthew 1902, pl. 1, fig. 9*a–b*), *Bradorona perspicator magna* (ROM 92CM; incomplete right valve; herein designated lectotype; Matthew 1902, pl. 1, fig. 11*a–b*), and *Bradorona perspicator major* (ROM 131CM; incomplete right valve; herein designated lectotype; Matthew 1902, pl. 1, fig. 10*a–b*), from divisions E1c, E2b and E3f respectively at Dugald Brook. *Bradorona spectator acuta* (syntypes: ROM 76CM A, largely incomplete valve, herein designated lectotype, Ulrich and Bassler 1931*a*, pl. 3, fig. 3; ROM 76CM B, ROM 26291 C) from divison E1b at Dugald Brook and E1d at Boundary Brook; Ulrich and Bassler (1931*a*) referred the latter taxon, in part, to their new species *Bradoria subquadrata* and, in part, to *Bradoria* sp. *Bradorona spectator aequata* (ROM 551CM; part and counterpart of exfoliated, posterodorsally incomplete left valve; termed 'type' by Ulrich and Bassler 1931*a*; Matthew 1902, pl. 1, fig. 14*a–b*) from divison E3d at Dugald Brook.

The poorly preserved *Indiana ovalis* Matthew, 1902 (ROM 87CM) and *Indiana ovalis prima* Matthew, 1902 (ROM 687CM), both referred to *Bradoria* by Ulrich and Bassler (1931*a*), are best regarded as *nomina dubia*. Of three syntypes of *Aparchites*(?) *robustus* Matthew, 1898 from the Hanford Brook Formation, which Ulrich and Bassler (1931*a*) referred to *Bradoria robusta*, the specimen they designated as lectotype (ROM 724CM(A)) is indeterminate; the two additional valves (ROM 724CM(B) B–C) are conspecific with *Indiana secunda* (Matthew, 1895). We have not traced the types of either *Bradorona observator ligata* Matthew, 1902, from division E3e at Dugald

Brook, or *Leperditia*?? *rugosa* Matthew, 1902 from division E3f in the Dugald Formation at Indian Brook, two taxa which were referred to *Bradoria* and *Bradorona* species by Ulrich and Bassler (1931*a*).

We consider the following referred specimens to be conspecific with *B. scrutator*:
1. Specimens of Ulrich and Bassler 1931*a*. Eight specimens (*ex* USNM 50517) from Dugald Brook (divisions E2a–b), referred to *B. robusta*: large right valve (USNM 483181), three poorly preserved specimens (USNM 483182–483184), an incomplete open carapace, a closed carapace and a right valve (USNM 483189, 483891, 486352; Pl. 1, figs 12, 14); and a right valve (USNM 50518; Ulrich and Bassler 1931*a*, pl. 2, fig. 4). Two well-preserved valves referred to *B. scrutator* (USNM 50540). A poorly preserved left valve (USNM 50532), referred to *B. perspicator*. Two well-preserved juvenile left valves (USNM 50542) and two poorly preserved valves (USNM 483193–483194, *ex* USNM 50538), referred to *B. benepuncta*. Four carapaces, a left valve and a valve fragment (USNM 483885 and USNM 486354–486356, 486353, 483884, *ex*. USNM 50537; Pl. 1, figs 3–4), two left valves (USNM 50535–50536) and a left valve and carapace (USNM 483886–483887, *ex* USNM 50534), referred to *B. acuta*. A carapace (USNM 483190, *ex* USNM 50525), referred to *B. vigilans* (the two other specimens which Ulrich and Bassler 1931*a* referred to that taxon, USNM 483191–483192, are indeterminate fragments). An incomplete carapace (and an indeterminate valve; *ex* USNM 50515, renumbered USNM 483882–483883), referred to *B. rugulosa*.

Incomplete right and left valves (USNM 483195–483196, *ex* USNM 50528), which Ulrich and Bassler (1931*a*, p. 27) referred to *B. obesa*, are questionably referred to *B. scrutator*; another specimen which they questionably referred to *B. obesa* is indeterminate (USNM 50529).
2. A juvenile right valve (BGS GSM 51691) from Shropshire, England, which Cobbold (1934, p. 380) referred to *B. benepunctata* [*sic*]. This determination indicates a Mid, rather than Early Cambrian age for the *Acrothele prima* Shale (Rushton *et al.* in press; Williams and Siveter in press).

Occurrence. Early Mid Cambrian Dugald Formation (divisions E1b to E3f; see Text-fig. 3), Dugald

---

EXPLANATION OF PLATE 2

Fig. 1. *Bradoria*? cf. *Bradoria oculata* (Matthew, 1895); ROM 718CM; Early Cambrian Hanford Brook Formation, Hanford Brook, New Brunswick; partly exfoliated left valve (holotype of *Leperditia*? *minor* Matthew, 1895, pl. 8, fig. 4*a*–*b*), stereo-pair; ×10.

Fig. 2 *Bradoria*? *cambrica* (Matthew, 1895); ROM 115CM(A); Early Cambrian Hanford Brook Formation, Hanford Brook, New Brunswick; partly exfoliated left? valve, lectotype (Matthew 1895, pl. 7, fig. 10*a*–*b*); ×10.

Fig. 3. *Bradoria*? *oculata* (Matthew, 1895); ROM 717CM; Early Cambrian Hanford Brook Formation, Hanford Brook, New Brunswick; left valve of posteriorly incomplete carapace, holotype (Matthew 1895, pl. 8, fig. 2*a*–*b*), stereo-pair; ×10.

Figs 4–8, 10–11. *Walcottella apicalis* Ulrich and Bassler, 1931*a*; Mid Cambrian Bright Angel Shale, Bass Canyon (figs 4–7, 10–11) and Chuar Valley (fig. 8), Grand Canyon, Arizona. 4, USNM 486431; incomplete right valve; ×9. 5–7, USNM 56483; posterodorsally incomplete carapace (holotype of *Walcottella pulchella* Ulrich and Bassler, 1931*a*, pl. 4, figs 11–12); ×9. 5, dorsal view, stereo-pair. 6, left valve, stereo-pair. 7, right valve. 8, USNM 56478; partly exfoliated, mid-dorsally flattened, left valve, lectotype (Ulrich and Bassler 1931*a*, pl. 4, fig. 1), stereo-pair; ×9. 10, USNM 56477; posteriorly and posteroventrally incomplete right valve (paralectotype of *Walcottella apicalis* Ulrich and Bassler, 1931*a*, pl. 4, fig. 2); ×9. 11, USNM 56489; partly exfoliated, mid-anteriorly damaged right valve (holotype of *Walcottella ventrosa* Ulrich and Bassler, 1931*a*, pl. 5, figs 4–5); ×10.

Figs 9, 12. *Walcottella apicalis* Ulrich and Bassler, 1931*a*?; Mid Cambrian Bright Angel Shale, Bass Canyon (fig. 9) and Chuar Valley (fig. 12), Grand Canyon, Arizona. 9, USNM 56485, largely exfoliated, internal mould of right valve (holotype of *Walcottella nitida* Ulrich and Bassler, 1931*a*, pl. 4, fig. 9), stereo-pair; ×8. 12, USNM 486435; right valve of internal mould of carapace (lectotype of *Indiana faba* Ulrich and Bassler, 1931*a*, pl. 5, figs 10–11); ×9.

All are lateral views except fig. 5.

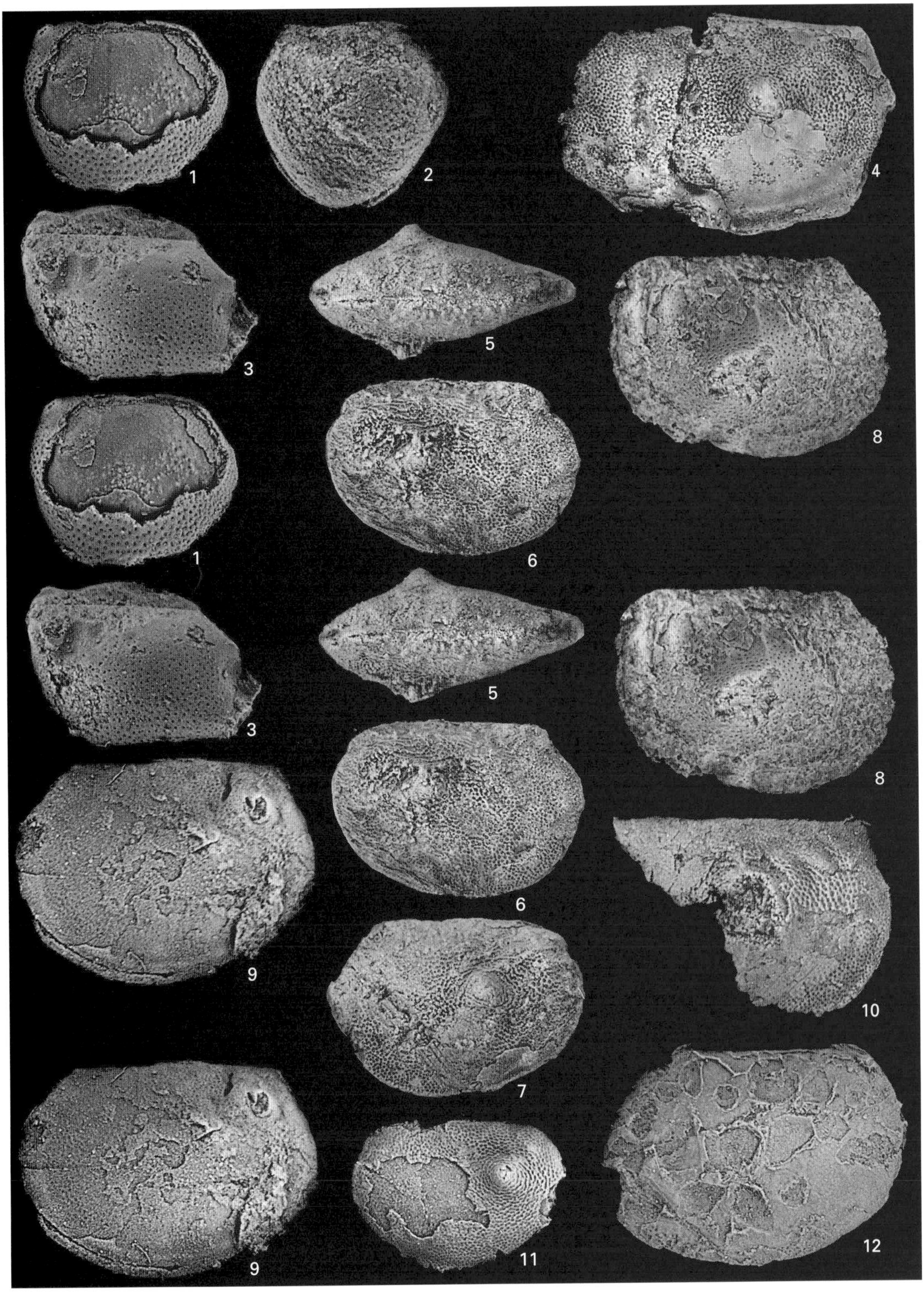

SIVETER and WILLIAMS, *Bradoria*?, *Walcottella*

Brook, Indian Brook and Boundary Brook, Cape Breton, Nova Scotia and early Mid Cambrian *Acrothele prima* Shale, Shropshire, England.

Bradoria? *cambrica* (Matthew, 1895)

Plate 2, figure 2

1895 *Schmidtella cambrica*, n. sp., Matthew, p. 137, pl. 7, fig. 10*a*–*b*.
1931*a* *Bradoria cambrica* (Matthew); Ulrich and Bassler, p. 28, pl. 1, fig. 10; pl. 3, figs 14–15.
1956 *Bradoria cambrica* (Matthew); Lochman, p. 1356.

Lectotype. Designated Ulrich and Bassler (1931*a*, p. 28); ROM 115CM (A) and (B) (Matthew 1895, pl. 7, fig. 10*a*–*b*; Pl. 2, fig. 2), part and counterpart of a partly exfoliated left? valve; Early Cambrian *Protolenus* trilobite Biozone, Hanford Brook Formation, Hanford Brook, New Brunswick.

Material and measurements. The lectotype (3·00 mm long) and an additional valve (USNM 50533; 2·50 mm long).

Remarks. This species differs from *B. scrutator* by its more rounded lateral outline and by lacking an anterodorsal node. Ulrich and Bassler's (1931*a*) material of *B*.? *cambrica* now comprises a single unfigured valve (USNM 50533): an empty vial with this specimen may have contained their figured specimens.

Occurrence. Type locality and stratum.

Bradoria? *oculata* (Matthew, 1895)

Plate 2, figure 3, ?figure 1

1895 *Primitia oculata*, n. sp., Matthew, p. 136, pl. 8, fig. 2*a*–*b*.
?1895 *Leperditia*(?) *minor*, n. sp., Matthew, p. 138, pl. 8, fig. 4*a*–*b*.
1899 *Primitia oculata*; Matthew, p. 204 [referred to *Bradoria*].
1902 *Primitia oculata*; Matthew, p. 452 [referred to *Bradoria*].
1931*a* *Bradoria oculata* (Matthew); Ulrich and Bassler, p. 16, pl. 1, figs 13–14 [original figures of Matthew].
?1931*a* *Bradoria minor* (Matthew); Ulrich and Bassler, p. 29, pl. 1, fig. 22.
1956 *Bradoria oculata* (Matthew); Lochman, p. 1356.
?1956 *Bradoria minor* (Matthew); Lochman, p. 1356.

Holotype. An incomplete carapace, ROM 717CM, 3·22 mm long (Matthew 1895, pl. 8, fig. 2*a*–*b*; Pl. 2, fig. 3); Early Cambrian *Protolenus* trilobite Biozone, Hanford Brook Formation, Hanford Brook, New Brunswick.

Remarks. The only known specimen of *B*.? *oculata* is a partly exfoliated, posteriorly incomplete carapace. It has a small anterodorsal node and puncta. In lateral valve outline it lacks the distinctive anterior curve of *B. scrutator*. It is also older than, and from a location different from that of *B. scrutator*.

Leperditia? *minor* Matthew, 1895, referred to *Bradoria* by Ulrich and Bassler (1931*a*) and known from a single partly exfoliated left valve (ROM 718CM, holotype; Pl. 2, fig. 1), is similar in size to *B*.? *oculata*. It also has a small anterodorsal node and puncta and is from the type Hanford Brook Formation. It may be synonymous with *B*.? *oculata*.

Occurrence. Type locality and stratum.

Bradoria sp. indet.

1963 *Bradoria* sp., Bolton and Copeland, p. 1070, pl. 143, figs 5–7.
1993*b* *Bradoria* sp. Bolton and Copeland; Hinz-Schallreuter, p. 413.

Material and measurements. A carapace, GSC 14592 (3·20 mm long).

Remarks. In valve lateral outline and ornament this specimen is like *B. scrutator*, from which it differs from valves of similar size by apparently lacking an anterodorsal node.

Occurrence. Mid Cambrian *Albertella* trilobite Biozone shales, Windsor Mountain, south-western Alberta.

Genus INDIANA Matthew, 1902

1902 *Indiana*, n. gen., Matthew, pp. 460–461 [*partim*], *non Primitia*(?) *fusiformis* (Matthew, 1895) [= *Mononotella* Ulrich and Bassler, 1931*a*] and *Indiana ovalis* Matthew, 1902 [= *nomen dubium*].
1903 *Indiana*; Matthew, pp. 169–170 [*partim*], *non Primitia*(?) *fusiformis* (Matthew, 1895) [= *Mononotella* Ulrich and Bassler, 1931*a*] and *Indiana ovalis* Matthew, 1902 [= *nomen dubium*].
1931*a* *Indiana* Matthew; Ulrich and Bassler, p. 68.
1931*b* *Indianites* new name, Ulrich and Bassler, p. 364.
1945 *Indianites* Ulrich and Bassler, 1931; Resser, p. 217.
1961 *Indiana* Matthew, 1902; Sylvester-Bradley, p. Q103.
1961 *Indianites* Ulrich and Bassler, 1931; Sylvester-Bradley, p. Q103.
1968 *Indianites*; Öpik, p. 18.
1968 *Indiana* Matthew; Öpik, p. 18.

Type species. Subsequently designated Ulrich and Bassler (1931*a*, p. 68); *Indiana lippa* Matthew, 1902, from the Middle Cambrian of Nova Scotia.

Diagnosis. Medium to large (adults *c.* 10–14 mm long), elongate postplete bradoriids. Carapace length generally one-third greater than height. Hinge-line between one-half and two-thirds valve length. Lateral outline of non dorsal parts of valve is gently curved. No lobation, nodation or marginal structures; in transverse section valves are gently and evenly convex. Valves have concentric granulostriation, punctation or are smooth.

Remarks. Indiana differs from *Bradoria* by its more elongate, postplete outline and by always lacking an anterodorsal node (see discussion of Bradoriidae above). Some species of *Indiana* have a phosphatic carapace (Siveter and Williams 1995).

Indiana impressa Ulrich and Bassler, 1931*a* (e.g. pl. 10, fig. 2; Resser 1945, pl. 27, fig. 7), from the Mid Cambrian Bright Angel Shale of Arizona, is based on several exfoliated and incomplete juveniles (USNM 56463), which are possibly referable to *Dielymella*. *Indiana curta* Ulrich and Bassler, 1931*a* (pl. 5, fig. 9; Resser 1945, pl. 26, fig. 15), also from the Bright Angel Shale, is based on an indeterminate incomplete valve (USNM 56466). The surviving specimen of *Indiana labiosa* Ulrich and Bassler, 1931*a* (USNM 56464 = pl. 9, fig. 26; specimen of pl. 9, fig. 27 is apparently lost), from the Mid Cambrian Stephen Formation of British Columbia, is a severly abraded, indeterminate valve. *Primitia acadica* Matthew, 1886 (p. 66, pl. 6, fig. 22, 22*a–b*), from the Mid Cambrian Fossil Brook Member of the Chamberlains Brook Formation at Porter Brook, New Brunswick, which was referred to *Indiana*? by Ulrich and Bassler (1931*a*, p. 82, pl. 9, fig. 13), is an anteriorly incomplete, largely exfoliated, indeterminate valve (ROM 619CM).

We have not examined the single valve (USNM 92725), from the Early Cambrian Kinzers Formation of Pennsylvania, which Resser and Howell (1938) assigned to *Indianites*? sp., or the type and only specimen of *Leperditia harrisi* Frederickson, 1946 (holotype, OUMIP 105–17F-1), from the Upper Cambrian of Oklahoma. Their relationships to *Indiana* need investigation.

Occurrence. Lower Cambrian of New Brunswick (Ulrich and Bassler 1931*a*) and lower Middle Cambrian of Nova Scotia (Matthew 1902); Lower Cambrian of New York State (Walcott 1887), southern Britain (Siveter and Williams 1995) and Scandinavia (Wiman 1905).

Indiana lippa Matthew, 1902

Plate 3, figures 2–4

1902 *Indiana lippa*, n. sp., Matthew, p. 463, pl. 2, fig. 10*a–d*.
1903 *Indiana lippa*; Matthew, p. 171, pl. 13, fig. 10*a–d*.

1931*a* *Indiana lippa* Matthew; Ulrich and Bassler, p. 70, pl. 9, figs 1–4.
?1931*a* *Indiana matthewi*, new species, Ulrich and Bassler, p. 76, pl. 9, fig. 15.
?1931*a* *Indiana primaeva* (Matthew); Ulrich and Bassler, p. 78 [*partim*; specimens from Hastings Cove?], *non* pl. 9, figs 9 [= *I. secunda*] and 10 [= *I. secunda*?].
1956 *Indiana lippa* (Matthew) [*sic*]; Lochman, p. 1356.
1961 *I. lippa*; Sylvester-Bradley, p. Q103, fig. 39.3*a–c*.

Lectotype. Subsequently designated by Ulrich and Bassler (1931*a*, p. 70); ROM 162CM (Matthew 1902, pl. 2, fig. 10*a–d*; Pl. 3, figs 2–4), a partly exfoliated carapace; early Mid Cambrian Dugald Formation (division E3f; see Text-fig. 3), Bourinot Group, Dugald Brook, Cape Breton, Nova Scotia.

Material. Three specimens: ROM 162CM, USNM 56474–56475. Possibly conspecific are USNM 50555 and USNM 56465.

Diagnosis. Moderately postplete, smooth species of *Indiana*; anteroventral part of valve is not flattened.

Description. Smooth, postplete, equivalved carapace, lacking posterior gape. Dorsal margin straight, remaining part of lateral outline convex with no marked curves. Hinge-line well developed, about two-thirds of valve length. Commissure straight. No lobation, nodation or marginal structures.

Remarks. *I. lippa* resembles *Indiana lentiformis* (Cobbold, 1921) from the Lower Cambrian of Britain (see Siveter and Williams 1995), but has a more rounded lateral outline posteriorly, its dorsal lateral outline is straighter and it lacks flattening of its valves anteroventrally. *I. lippa* differs from the early Cambrian *Indiana secunda* (Matthew, 1895) by being smooth and much less postplete.

Four poorly preserved, juvenile valves (USNM 50555), from Mid Cambrian limestones (Fossil Brook Member of the Chamberlains Brook Formation?) at Hastings Cove, New Brunswick, which are the syntypes of *Indiana matthewi* Ulrich and Bassler, 1931*a*, are possibly conspecific with *I. lippa*, as are their specimens (USNM 56465) from Hastings Cove, which they referred to *I. primaeva* (Matthew).

Occurrence. Early Mid Cambrian Dugald Formation (divisions E3e, E3f; see Text-fig. 3), Bourinot Group, Dugald Brook, Cape Breton, Nova Scotia. Possibly from Mid Cambrian limestones at Hastings Cove, New Brunswick.

EXPLANATION OF PLATE 3

Fig. 1. *Walcottella apicalis* Ulrich and Bassler, 1931*a*?; USNM 56484; Mid Cambrian Bright Angel Shale, Bass Canyon, Grand Canyon, Arizona; damaged, partly exfoliated left valve (holotype of *Walcottella leperditoides* Ulrich and Bassler, 1931*a*, pl. 4, fig. 8), stereo-pair; ×9.

Figs 2–4. *Indiana lippa* Matthew, 1902; Mid Cambrian Dugald Formation, Dugald Brook, Cape Breton, Nova Scotia; ROM 162CM; partly exfoliated carapace, lectotype (Matthew 1902, pl. 2, fig. 10*a–d*); ×9. 2, left valve, stereo-pair; 3, dorsal view, stereo-pair; 4, right valve.

Figs 5–10. *Indiana secunda* (Matthew, 1895); Early Cambrian Hanford Brook Formation, Long Island (figs 5–6) and Hanford Brook (figs 7–10), New Brunswick. 5, ROM 169CM A; exfoliated left valve (lectotype of *Primitia pyriformis* Matthew, 1898, pl. 1, fig. 3*a–c*); ×8. 6, ROM 724CM(B) C; silicon rubber cast of external mould of right valve; ×8. 7, ROM 53CM; mostly exfoliated, anteriorly incomplete left valve (holotype of *Leperditia*(?) *primaeva* Matthew, 1895, pl. 8, fig. 6*a–b*); ×8. 8–10, ROM 49CM; mostly exfoliated carapace, holotype (Matthew 1895, pl. 7, fig. 11*a–b*); ×9. 8, ventral view. 9, dorsal view, stereo-pair. 10, right valve, stereo-pair.

All are lateral views except figs 3, 8–9.

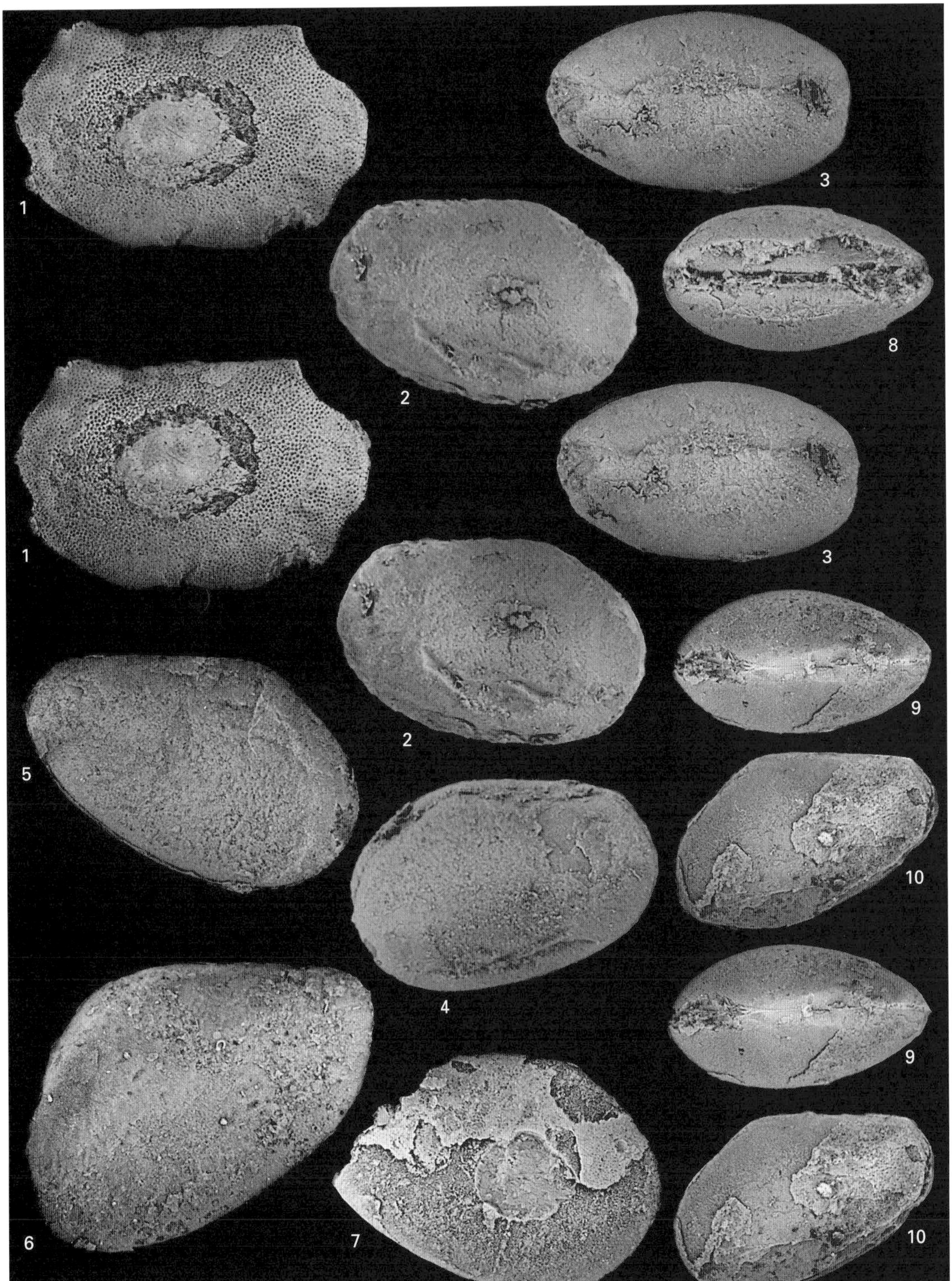

SIVETER and WILLIAMS, *Walcottella, Indiana*

Indiana secunda (Matthew, 1895)

Plate 3, figures 5–10

1895 *Aparchites secunda*, n. sp., Matthew, p. 136, pl. 7, fig. 11*a–b*.
1895 *Leperditia*(?) *primaeva*, n. sp., Matthew, p. 138, pl. 8, fig. 6*a–b*.
1898 *Primitia pyriformis*, n. sp., Matthew, p. 132, pl. 1, fig. 3*a–c*.
1898 *Aparchites*(?) *robustus*, n. sp., Matthew, p. 132 [*partim*], *non* pl. 1, fig. 4*a–c* [= indeterminate].
1902 *Primitia pyriformis*; Matthew, p. 461.
1902 *Aparchites*(?) *robustus*; Matthew, p. 461.
1906 *Bradorna*(?) *robusta* [*sic*]; Matthew, p. 406.
1931*a* *Indiana secunda* (Matthew); Ulrich and Bassler, p. 71, pl. 9, figs 5–8.
1931*a* *Indiana pyriformis* (Matthew); Ulrich and Bassler, p. 72, pl. 9, figs 11–12.
1931*a* *Indiana longula*, new species, Ulrich and Bassler, p. 73, pl. 9, figs 19–21.
1931*a* *Indiana primaeva* (Matthew); Ulrich and Bassler, pp. 77–78 [*partim*], pl. 9, fig. 9; ?fig. 10 [*non* specimens from Hastings Cove = *Indiana lippa*?].
?1932 *Bradoria* sp., Poulsen, p. 24, pl. 4, figs 8–10.
*non*1936 *Indianites* cf. *primaeva* (Matthew); Cobbold, p. 228, pl. 14, fig. 9 [= indeterminate].
*non*1936 *Indianites* cf. *pyriformis* (Matthew); Cobbold, p. 229, pl. 15, figs 5–6 [= indeterminate].
1956 *Indiana secunda* Matthew [*sic*]; Lochman, p. 1356.

Holotype. A largely exfoliated juvenile carapace, ROM 49CM (type by monotypy; Matthew 1895, p. 136, pl. 7, fig. 11*a–b*; Pl. 3, figs 8–10); Early Cambrian Hanford Brook Formation, Hanford Brook, New Brunswick.

Material. Sixteen specimens: ROM 49CM; 53CM; 169CM A–B; 724CM(B) B–C. USNM 50558 and 56470; NBMG 3306, 9893–9895, 9897; OUM AT.112–AT.113. NBMG 8058–8060 and 8098–8099 may also contain other material but were not available for study by us. We have not located the additional ROM specimens mentioned by Ulrich and Bassler (1931*a*, p. 71).

Diagnosis. Markedly postplete, punctate species of *Indiana*; anteroventral part of valve is not flattened.

Description. Equivalved, markedly postplete carapace, lacking posterior gape. Commissure straight. Dorsal margin straight, hinge-line well developed and about two-thirds of valve length. Remaining lateral outline convex with no curves. No lobation, nodation or marginal structures. Ornament consists of fine, closely spaced puncta.

Remarks. This species differs from the Mid Cambrian *I. lippa* in being punctate and more postplete. We consider that *Leperditia*(?) *primaeva* Matthew, 1895, *Primitia pyriformis* Matthew, 1898, and *Indiana longula* Ulrich and Bassler, 1931*a* are junior synonyms of *I. secunda*. The holotype of *L*.(?) *primaeva* is a large, anteriorly incomplete left valve, ROM 53CM (type by monotypy; Matthew, 1895, pl. 8, fig. 6*a–b*; Pl. 3, fig. 7). Syntypes of *P. pyriformis* are two large left valves, ROM 169CM B and ROM 169CM A (Matthew 1898, pl. 1, fig. 3*a–c*; Pl. 3, fig. 5; designated lectotype by Ulrich and Bassler, 1931*a*, p. 122). The holotype of *I. longula* is a distorted carapace (USNM 50558). An incomplete right valve (USNM 56470), which Ulrich and Bassler (1931*a*) referred to *Indiana pyriformis*, is also conspecific with *I. secunda*.

Of three syntypes of *Aparchites*(?) *robustus* Matthew, 1898, the lectotype designated by Ulrich and Bassler (1931*a*) is an indeterminate external mould (ROM 724CM(A)). The two other specimens, ROM 724CM(B) B (external mould of a large left valve) and ROM 724CM(B) C (external mould of a large right valve; see Pl. 3, fig. 6) are conspecific with *Indiana secunda*. The carapace referred to *Bradoria* sp. by Poulsen (1932), from the Early Cambrian Bastion Formation of East Greenland, is possibly conspecific with *I. secunda*.

The Early Cambrian *Indiana dermatoides* (Walcott, 1887) apparently differs from *I. secunda* by its shorter hinge. Specimens reported as co-occurring with *I. dermatoides* from Washington County, New York State and referred to *I. secunda* and *I. pyriformis* (see Ulrich and Bassler 1931*a*, p. 72) are probably incomplete specimens of *I. dermatoides*. *Indianites* cf. *primaeva* and *Indianites* cf. *pyriformis* of Cobbold (1936), from the Early Cambrian Lower Comley Limestones of Shropshire, England, are indeterminate (Williams and Siveter in press).

Occurrence. Early Cambrian *Protolenus* trilobite Biozone, Hanford Brook Formation at Hanford Brook and Long Island, New Brunswick. Possibly from the Lower Cambrian of East Greenland (Poulsen 1932).

Indiana dermatoides (Walcott, 1887)

Plate 4, figures 6–8

1887 *Leperditia dermatoides*, n. sp., Walcott, p. 192, pl. 1, fig. 13, 13*a*.
1890 *Leperditia dermatoides* Walcott; Walcott, p. 626, pl. 80, fig. 1, 1*a*.
1891 *Leperditia dermatoides*; Matthew, p. 290.
1918 *Indiana dermatoides* (Walcott); Chapman, p. 109.
non 1921 *Leperditia*(?) *dermatoides* Walcott(?); Cobbold, p. 368 [= *Dielymella*?].
1929 *Indiana dermatoides*; Gürich, p. 35, text-fig. 1.4 [on p. 37 referred to as *Leperditia dermatoides* Walcott, 1887].
1931*a* *Indiana dermatoides* (Walcott); Ulrich and Bassler, p. 75, pl. 9, figs 16–17.
1931*a* *Indiana suberecta*, new species, Ulrich and Bassler, p. 74, pl. 9, fig. 28.
1931 *Indiana dermatoides* Walc.; Kummerow, pp. 244, 252, fig. 12.
1956 *Indiana dermatoides* (Walcott); Lochman, p. 1376, pl. 4, figs 7–8.

Lectotype. Designated Ulrich and Bassler (1931*a*, p. 75); USNM 17448 (Walcott 1887, pl. 1, fig. 13, 13*a*; Pl. 4, fig. 8), a partly exfoliated left valve; early Cambrian limestones interbedded with shales in Washington County, New York State.

Material. Twenty-two specimens: USNM 17474, 17448, 56469, 126755.

Measurements. Most specimens are incomplete. Lochman's (1956) figured specimen (USNM 126755) and the lectotype are 5·24 mm long.

Diagnosis. Markedly postplete, punctate species of *Indiana* with hinge-line shorter than half valve length.

Description. Markedly postplete carapace with short hinge-line. Lateral outline of free margin convex with no marked curves. No lobation, nodation or marginal structures. Fine, closely spaced puncta.

Remarks. *Indiana dermatoides* differs from the Mid Cambrian *I. lippa* by being more postplete and by having a shorter hinge, and from *I. secunda* by its shorter hinge. However, the dorsal margin of the lectotype of *I. dermatoides* is partially crushed and thus, as preserved, the hinge is not fully representative.

Ulrich and Bassler (1931*a*, p. 76, pl. 9, figs 16–17) refigured Walcott's original specimen but wrongly referred to it as USNM 17474, which is a collection of about 20, less well preserved specimens of *I. dermatoides* from Washington County, New York State. We consider *Indiana suberecta* Ulrich and Bassler, 1931*a* (USNM 56469, holotype, an incomplete right? valve; Pl. 4, fig. 7), from Washington County, to be synonymous with *I. dermatoides*.

Occurrence. Early Cambrian *Callavia–Protolenus* trilobite biozones, limestones interbedded with shales from several small localities in the vicinity of Greenwich, Washington County and in the *Ellipsocephala asaphoides* fauna of the Cambridge and Hoosick quadrangles, New York State (Walcott 1887, p. 193; Ulrich and Bassler 1931*a*, p. 76; Lochman 1956).

Genus INDOTA Öpik, 1968

1968 *Indota* nov., Öpik, p. 18.
?1968 *Tropidiana* nov., Öpik, p. 20.

Type species. By original designation; *Indota otica* Öpik, 1968, from the Lower Cambrian of Queensland, Australia.

Diagnosis. Medium to large (adults *c*. 10 mm long), elongate, postplete bradoriids having a lateral furrow which is sited some distance from and parallel to the free margin of the valve. Dorsal margin straight, about two-thirds of valve length. Valves granulose or smooth.

Remarks. Like *Indiana*, *Indota* (and the probably synonymous *Tropidiana* Öpik, 1968) from the Early Cambrian Yelvertoft Beds of Queensland, Australia (Öpik 1968), has an elongate, postplete lateral shape and lacks nodes. *Indota* differs by having a lateral furrow some distance from and parallel to the free margin of the valve. This feature possibly corresponds to the inner margin of a relatively broad doublure (see Öpik 1968, p. 21).

Occurrence. Lower Cambrian of Australia (Öpik 1968) and the USA (herein). Possibly from the Lower Cambrian of Kazakhstan (Melnikova *et al*. in press).

Indota tennesseensis (Resser, 1938)

Plate 4, figures 4–5

1938 *Indianites tennesseensis*, n. sp., Resser, p. 107, pl. 3, fig. 47.
1963 *Indiana tennesseensis* (Resser); Laurence and Palmer, p. 54, fig. 73.2*a–b*.
1968 '*Indianites*' *tennesseensis* Resser, 1938; Öpik, p. 18 [referred to *Indota*].
?1968 *Indota otica* sp. nov., Öpik, p. 19, pl. 1, figs 1–4; text-fig. 1.
?1968 *Tropidiana cirrata* sp. nov., Öpik, p. 20, pl. 1, figs 5–6; text-figs 2, 11.

Holotype. An anterodorsally incomplete right valve and counterpart, USNM 94759 (Resser 1938, pl. 3, fig. 47; Pl. 4, fig. 4); Early Cambrian Murray Shale at Little River Gap, Chilhowee Mountain, Tennessee; collected by Walcott and Keith in 1893.

Material. Three well-preserved valves (USNM 94759, 143799–143800) and 14 incomplete carapaces, valves and fragments (USNM 208751–208764).

EXPLANATION OF PLATE 4

Figs 1–3. *Mononotella fusiformis* (Matthew, 1895); Early Cambrian Hanford Brook Formation, Hanford Brook, New Brunswick. 1–2, ROM 132CM; right half of carapace, holotype (Matthew 1895, pl. 8, fig. 3*a–b*); ×8. 1, lateral view, stereo-pair. 2, dorsal view (anterior to the left). 3, USNM 56499; partly exfoliated, anteriorly damaged carapace (Ulrich and Bassler 1931*a*, pl. 8, fig. 25), dorsal view; ×8.

Figs 4–5. *Indota tennesseensis* (Resser, 1938); Early Cambrian Murray Shale, Little River Gap, Chilhowee Mountain, Tennessee. 4, USNM 94759; internal mould of anterodorsally incomplete right valve, holotype (Resser 1938, pl. 3, fig. 47), stereo-pair; ×8. 5, USNM 143799; dorsally damaged left valve (Laurence and Palmer 1963, fig. 73.2A); ×8.

Figs 6–8. *Indiana dermatoides* (Walcott, 1887); Early Cambrian, New York State: *c*. 1 km west of Cambridge (fig. 6); vicinity of Greenwich, Washington County (figs 7–8). 6, USNM 126755; largely exfoliated right valve (Lochman 1956, pl. 4, figs 7–8); ×8·5. 7, USNM 56469; partly exfoliated, anteriorly incomplete ?right valve (holotype of *Indiana suberecta* Ulrich and Bassler, 1931*a*, pl. 9, fig. 28); ×8. 8, USNM 17448; partly exfoliated left valve, lectotype (Walcott 1887, pl. 1, fig. 13, 13*a*), stereo-pair; ×8.

Fig. 9. *Beyrichona papilio* Matthew, 1886; ROM 114CM; Early Cambrian Hanford Brook Formation, Hanford Brook, New Brunswick; internal mould of flattened, open carapace, holotype (Matthew 1886, pl. 6, fig. 20, 20*a–b*); ×10.

Fig. 10. *Beyrichona papilio* Matthew, 1886?; USNM 483174; Early Cambrian Hanford Brook Formation, Hanford Brook, New Brunswick; posterodorsally incomplete right valve (Ulrich and Bassler 1931*a*, pl. 6, fig. 11), stereo-pair; ×9.

All are lateral views except figs 2–3.

SIVETER and WILLIAMS, bradoriids

Measurements. The holotype is 6·18 mm long. Other specimens are up to 10 mm long (Laurence and Palmer 1963).

Diagnosis. Species of *Indota* in which the admarginal part of the valve (that area demarcated by a furrow parallel to the free margin) is broad, measuring one-quarter of valve height.

Description. Elongate, postplete carapace. Dorsal margin straight, about two-thirds of valve length. Free margin part of lateral outline gently convex. Valve surface lacks lobes or nodes. Admarginal part of valve broad, demarcated from remainder of lateral valve surface by a narrow furrow, entire between cardinal corners, only faintly developed adjacent to dorsal margin. Valves granulose or smooth.

Remarks. Based on figures alone (Öpik 1968) the type species, *I. otica* from Australia, shows no fundamental difference in size or morphology from *I. tennesseensis*. Because of their wide palaeogeographical separation (see McKerrow *et al.* 1992) and the problems of homeomorphy in morphologically simple carapaces, *I. otica* is questionably placed in synonymy with *I. tennesseensis*. Differences between the type species, *Tropidiana cirrata* Öpik, 1968, and *I. otica* appear to be preservational; both are from the same Early Cambrian locality.

Occurrence. Early Cambrian Murray Shale, Chilhowee Group, at Little River Gap and Murray Gap, Chilhowee Mountain, Blount County, Tennessee.

Genus WALCOTTELLA Ulrich and Bassler, 1931*a*

Type species. By original designation; *Walcottella apicalis* Ulrich and Bassler, 1931*a*, from the Middle Cambrian of the Grand Canyon, Arizona.

Diagnosis. Medium-sized (adults *c*. 6–7 mm long), postplete to subamplete bradoriids with well-developed, acuminate anterocentral node around which are concentric reticulostriae. Anterodorsal node weakly developed.

Remarks. The late Cambrian Scandinavian phosphatocopid *Waldoria* Gründel, 1981 superficially resembles *Walcottella* in having a pronounced median node. The type species, *Waldoria buchholzi* Gründel, 1981, differs in having three subdorsally aligned nodes, a feature typical of vestrogothiid phosphatocopids (see Hinz-Schallreuter 1993*b*, p. 402).

We have not examined Rushton's (1978, pl. 24, fig. 9) Late Cambrian Manuels River specimen referred to as *Walcottella* (on rock slab with GSC 13055), but this may be a vestrogothiid.

Occurrence. Mid Cambrian *Glossopleura* trilobite Biozone, Arizona, USA (Ulrich and Bassler 1931*a*).

Walcottella apicalis Ulrich and Bassler, 1931*a*

Plate 2, figures 4–11, ?figure 12; ?Plate 3, figure 1

1931*a* *Walcottella apicalis*, new species, Ulrich and Bassler, p. 33, pl. 4, figs 1–2.
1931*a* *Walcottella concentrica*, new species, Ulrich and Bassler, p. 34, pl. 4, figs 3–4.
1931*a* *Walcottella breviuscula*, new species, Ulrich and Bassler, p. 34, pl. 4, figs 5–6.
1931*a* *Walcottella scitula*, new species, Ulrich and Bassler, p. 35, pl. 4, fig. 7.
1931*a* *Walcottella pulchella*, new species, Ulrich and Bassler, p. 36, pl. 4, figs 11–12.
?1931*a* *Walcottella leperditoides*, new species, Ulrich and Bassler, p. 37, pl. 4, fig. 8.
?1931*a* *Walcottella nitida*, new species, Ulrich and Bassler, p. 37, pl. 4, figs 9–10.
1931*a* *Walcottella oblonga*, new species, Ulrich and Bassler, p. 38, pl. 5, fig. 8.
?1931*a* *Walcottella obsoleta*, new species, Ulrich and Bassler, p. 39, pl. 4, fig. 13.
1931*a* *Walcottella limatula*, new species, Ulrich and Bassler, p. 40, pl. 5, figs 1–3.
1931*a* *Walcottella ventrosa*, new species, Ulrich and Bassler, p. 41, pl. 5, figs 4–5.

1931*a* *Walcottella subtruncata*, new species, Ulrich and Bassler, p. 41, pl. 5, fig. 6.
1931*a* *Walcottella longula*, new species, Ulrich and Bassler, p. 42, pl. 5, fig. 7.
?1931*a* *Indiana faba*, new species, Ulrich and Bassler, p. 79, pl. 5, figs 10–14.
?1945 *Indianites faba* (Ulrich and Bassler); Resser, p. 217, pl. 27, figs 9, 17.
1945 *Walcottella apicalis* Ulrich and Bassler; Resser, p. 218, pl. 26, fig. 18; pl. 27, fig. 1.
1945 *Walcottella breviuscula* Ulrich and Bassler; Resser, p. 218, pl. 27, figs 5, 22.
1945 *Walcottella concentrica* Ulrich and Bassler; Resser, p. 218, pl. 26, figs 12, 17.
?1945 *Walcottella leperditoides* Ulrich and Bassler; Resser, p. 218, pl. 27, fig. 3.
1945 *Walcottella limatula* Ulrich and Bassler; Resser, p. 218, pl. 27, figs 18, 25.
1945 *Walcottella longula* Ulrich and Bassler; Resser, p. 218, pl. 27, fig. 14.
?1945 *Walcottella nitida* Ulrich and Bassler; Resser, p. 219, pl. 27, fig. 16.
1945 *Walcottella oblonga* Ulrich and Bassler; Resser, p. 219, pl. 26, fig. 14.
?1945 *Walcottella obsoleta* Ulrich and Bassler; Resser, p. 219, pl. 27, fig. 2.
1945 *Walcottella pulchella* Ulrich and Bassler; Resser, p. 219, pl. 26, fig. 16.
1945 *Walcottella scitula* Ulrich and Bassler; Resser, p. 219, pl. 27, fig. 6.
1945 *Walcottella subtruncata* Ulrich and Bassler; Resser, p. 219, pl. 27, fig. 8.
1945 *Walcottella ventrosa* Ulrich and Bassler; Resser, p. 220, pl. 26, fig. 19.
1961 *W. apicalis*; Sylvester-Bradley, p. Q102, fig. 37.2*a*.
1961 *W. concentrica*; Sylvester-Bradley, p. Q102, fig. 37.2*b–c*.

Lectotype. Designated herein; a left valve, USNM 56478 (Ulrich and Bassler 1931*a*, pl. 4, fig. 1; Resser 1945, pl. 27, fig. 1; Pl. 2, fig. 8); sandstone bed in the Bright Angel Shale, Mid Cambrian *Glossopleura* trilobite Biozone, Chuar Valley, Grand Canyon, Arizona.

Material. Thirty-four, mostly incomplete specimens: USNM 56477–56483, 56486, 56488–56491, 486426 and 486429–486434. Possibly conspecific are USNM 56458–56459, 56484–56485, 56487, 486427–486428 and 486435–486437.

Measurements. Specimens from the Bright Angel Shale, Grand Canyon, are 2·5–6·5 mm long.

Diagnosis. As for the genus.

Description. Postplete to subamplete equivalved carapace, lacking posterior gape. Dorsal margin straight, hinge-line well developed, shorter than valve length. Commissure straight. Lateral outline of free margin evenly convex. Circular, anterocentral acuminate node, around which are concentric reticulostriae. Anterodorsal node weakly developed. Admarginal ridge narrow, entire between cardinal corners.

Remarks. Walcottella apicalis shows variation of its lateral valve outline, length-height ratio and, in particular, in the development and position of its anterocentral and anterodorsal nodes. These, and particularly preservational factors, account largely for the many 'species' of *Walcottella* recognized by Ulrich and Bassler (1931*a*), but which we regard as synonyms. Particular variants do not cluster at specific horizons or localities and, moreover, Ulrich and Bassler (1931*a*, pp. 33, 42) themselves consistently refer to the co-occurrence of their species in the Bright Angel Shale. Well preserved *W. apicalis* are reticulostriate but, where eroded, shells are punctate or smooth (cf. *Bradoria*). Of Ulrich and Bassler's species some were based on internal moulds (e.g. *W. nitida*) or single specimens (e.g. *W. scitula*, *W. pulchella*) and in others the acuminate node appears to be abraded away (e.g. *W. leperditoides*).

Syntypes of *W. apicalis* comprise the lectotype (USNM 56478), an incomplete, partly exfoliated, mid-dorsally flattened left valve (USNM 486426, *ex* USNM 56478), an incomplete right valve (USNM 56477; Ulrich and Bassler 1931*a*, pl. 4, fig. 2; Resser 1945, pl. 26, fig. 18; Pl. 2, fig. 10), and two poorly preserved, possibly conspecific specimens (USNM 486427–486428, *ex* USNM 56477).

We consider *W. concentrica*, *W. breviuscula*, *W. scitula*, *W. pulchella*, *W. oblonga*, *W. limatula*, *W. subtruncata*, *W. ventrosa* and *W. longula* (all Ulrich and Bassler, 1931*a*) to be synonyms of *W. apicalis*. Syntypes of *W. concentrica* are an incomplete right valve (USNM 56480; Ulrich and Bassler 1931*a*, pl. 4, fig. 3; Resser 1945, pl. 26, fig. 17; herein designated lectotype), an incomplete right? valve (USNM 486429, *ex* USNM 56479; Ulrich and Bassler 1931*a*, pl. 4, fig. 4; Resser 1945, pl. 26, fig. 12) and several incomplete, unfigured specimens (USNM 56479). Syntypes of *W. breviuscula* are a right valve (USNM 56481; Ulrich and Bassler 1931*a*, pl. 4, fig. 5; Resser 1945, pl. 27, fig. 22; herein designated lectotype) and a left valve (USNM 486430, *ex* USNM 56481; Ulrich and Bassler 1931*a*, pl. 4, fig. 6; Resser 1945, pl. 27, fig. 5). *W. scitula* was based on a posteriorly and anteroventrally incomplete juvenile left valve (holotype, USNM 56482; Ulrich and Bassler 1931*a*, pl. 4, fig. 7; Resser 1945, pl. 27, fig. 6) and *W. pulchella* on a partially exfoliated juvenile carapace (holotype, USNM 56483; Ulrich and Bassler 1931*a*, pl. 4, figs 11–12; Resser 1945, pl. 26, fig. 16; Pl. 2, figs 5–7).

Specimens of *W. oblonga* comprise a left valve (holotype, USNM 56486; Ulrich and Bassler 1931*a*, pl. 5, fig. 8; Resser 1945, pl. 26, fig. 14) and a less complete right valve (USNM 486431, *ex* USNM 56486; Pl. 2, fig. 4). Syntypes of *W. limatula* comprise a right valve (USNM 486432, *ex* USNM 56488; Ulrich and Bassler 1931*a*, pl. 5, fig. 1; Resser 1945, pl. 27, fig. 25), a carapace (holotype, USNM 486433, *ex* USNM 56488; Ulrich and Bassler 1931*a*, pl. 5, figs 2–3; Resser 1945, pl. 27, fig. 18) and several unfigured specimens (USNM 56488). Specimens of *W. subtruncata* include a right valve (holotype, USNM 56490; Ulrich and Bassler 1931*a*, pl. 5, fig. 6; Resser 1945, pl. 27, fig. 8) and an unfigured fragment (USNM 486434, *ex* USNM 56490). *W. ventrosa* (holotype, USNM 56489; Ulrich and Bassler 1931a, pl. 5, figs 4–5; Resser 1945, pl. 26, fig. 19; Pl. 2, fig. 11) and *W. longula* (holotype, USNM 56491; Ulrich and Bassler 1931*a*, pl. 5, fig. 7; Resser 1945, pl. 27, fig. 14) are both based on partly exfoliated right valves.

Possibly synonymous with *W. apicalis* are *W. nitida*, *W. leperditoides*, *W. obsoleta* and *Indiana faba* (all Ulrich and Bassler, 1931*a*). *W. nitida* is based on an exfoliated right valve (holotype, USNM 56485; Ulrich and Bassler 1931*a*, pl. 4, fig. 9; Resser 1945, pl. 27, fig. 16; Pl. 2, fig. 9), which has some puncta, an anterodorsal node and an anterocentral node which is more anterior than in typical *W. apicalis*. The holotypes of *W. leperditoides* (USNM 56484; Ulrich and Bassler 1931*a*, pl. 4, fig. 8; Resser 1945, pl. 27, fig. 3; Pl. 3, fig. 1) and *W. obsoleta* (USNM 56487; Ulrich and Bassler 1931*a*, pl. 4, fig. 13; Resser 1945, pl. 27, fig. 2) are large, partly exfoliated, possible adult specimens in which damage has removed the anterocentral node. Ulrich and Bassler's (1931*a*) illustration of *W. leperditoides* shows marked retouching, restoring the mostly incomplete free margin. Similarly, their illustration of *W. obsoleta* does not show the dorsal and ventral deformation of the valve which resulted in its markedly elongate shape. Syntypes of *Indiana faba* include three exfoliated specimens (USNM 486435, herein designated lectotype, Pl. 2, fig. 12, and USNM 486436–486437, all *ex* USNM 56458; Ulrich and Bassler 1931*a*, pl. 5, figs 10–15) which, although lacking ornament and nodation, are similar in size and lateral outline to the co-occurring *W. apicalis*. The holotype of *Indiana faba intermedia* Ulrich and Bassler, 1931*a* is a partially exfoliated valve which lacks evidence of an anterocentral node (USNM 56462; Ulrich and Bassler 1931*a*, pl. 5, fig. 15; Resser 1945, pl. 26, fig. 15); its systematic position is uncertain.

Occurrence. Mid Cambrian *Glossopleura* trilobite Biozone, Bright Angel Shale, Bass Canyon and Chuar Valley, Grand Canyon, Arizona.

Walcottella sp. indet.

1954 Leperditiid ostracode?; Palmer, p. 774, pl. 91, figs 14–15, 17.
1968 Leperditiid ostracode, Palmer; Öpik, p. 8 [tentatively referred to *Walcottella*].

Material and measurements. A right valve (UT 32198*b*, 3·0 mm long; number incorrectly given as UT 31189*b* in plate explanation of Palmer 1954) and a fragment (UT 32189*a*).

Remarks. Based on Palmer's (1954) two figured specimens, *Walcottella* sp. indet. has an acuminate lateral node but in a somewhat more anterior position than in *W. apicalis*. The presence of an anterodorsal node cannot be verified.

Occurrence. Late Cambrian *Aphelaspis* trilobite Biozone, Riley Formation, central Texas.

Family BEYRICHONIDAE Ulrich and Bassler, 1931*a*

Diagnosis. Small to medium-sized (adults *c*. 2·5–6 mm long), subamplete to slightly postplete, trilobate bradoriids having sub-triangular lateral outline, straight dorsal margin and well developed hinge-line. Equivalved carapace, lacking posterior gape. Anterior and posterior lobes narrow, anterior lobe normally has greater relief; central lobe broader and more subdued. All lobes confluent ventrally with lateral surface. Marginal ridge entire between cardinal corners in some species.

Remarks. Beyrichonids differ from hipponicharionids in having less inflated and less ventrally extended anterior and posterior lobes.

Hinz-Schallreuter (1993*b*) referred her genus *Tagra* and the monospecific *Uskutchiella* Melnikova, 1992 to Beyrichonidae. The holotype, and only specimen, of the type species *Tagra koppkai* Hinz-Schallreuter, 1993*b* is an abraded, posteriorly incomplete right valve whose lobal morphology resembles that of *Beyrichona triceps* Matthew, 1903. The holotype of the type species *Uskutchiella sulcata* Melnikova, 1992 (see Melnikova *et al*. in press), is a flattened, severely abraded, open carapace whose lateral outline and subdued lobation (as preserved), suggest a beyrichonid affinity.

Genus BEYRICHONA Matthew, 1886

Type species. Subsequently designated Ulrich and Bassler (1931*a*, p. 43); *Beyrichona papilio* Matthew, 1886, from the Lower Cambrian of New Brunswick.

Diagnosis. Small to medium-sized (adults *c*. 2·5–6·0 mm long), subamplete to postplete, sub-triangular bradoriids. Lateral outline is convex to acuminate posterior of midlength ventrally, anterodorsally is produced into a marked curve in some species. Weakly to markedly trilobate; lobes developed dorsally, ventrally confluent with lateral surface, separated by weak sulci. Marginal ridge narrow, entire between cardinal corners. Surface smooth.

Remarks. Rarely, in some specimens of *Beyrichona*, a series of anastomosing ridges is observed laterally on internal moulds (e.g. *Beyrichona gevalensis* Wiman, 1905; see Vannier *et al*. in press).

The smooth valve surface of *Beyrichona* contrasts with the granulose ornament of the hipponicharionid *Hipponicharion* Matthew, 1886 and the concentric striation typical of the Bradoriidae *Bradoria* and *Indiana*.

Occurrence. Lower Cambrian of New Brunswick (Matthew 1886) and Middle Cambrian of Nova Scotia (Matthew 1903). Also the ?Lower Cambrian of Scandinavia (Wiman 1905), the Middle Cambrian and lower Ordovician (Tremadoc) of southern Britain (Rushton *et al*. in press; Williams and Siveter in press), the Upper Cambrian of Belarus and the Middle and Upper Cambrian of Kazakhstan (Melnikova *et al*. in press).

Beyrichona papilio Matthew, 1886

Plate 4, figure 9, ?figure 10; Plate 5, figure 1

1886 *Beyrichona papilio*, n. sp., Matthew, p. 65, pl. 6, fig. 20, 20*a*–*b*.
1895 *Beyrichona papilio*; Matthew, p. 134, pl. 7, fig. 4*a*–*c*.
non 1895 *Beyrichona triangula* n. sp., Matthew, p. 135, pl. 7, fig. 5 [= indeterminate composite of more than one specimen].
1931*a* *Beyrichona papilio* Matthew; Ulrich and Bassler, p. 43 [*partim*, pl. 7, fig. 5, ?pl. 6, fig. 11, ?pl. 7, fig. 7 [specimen now lost]]; *non* pl. 7, fig. 6 [= indeterminate].

1956 *Beyrichona papilio* (Matthew) [*sic*]; Lochman, p. 1357.
1961 *B. papilio*; Sylvester-Bradley, p. Q102, fig. 38.1*a*.

Holotype. A flattened, open carapace and its counterpart, ROM 114CM (type by monotypy; Matthew 1886, pl. 6, fig. 20, 20*a–b*; Ulrich and Bassler 1931*a*, pl. 7, fig. 5; Pl. 4, fig. 9); Early Cambrian *Protolenus* trilobite Biozone, Hanford Brook Formation, Hanford Brook, New Brunswick.

Material and measurements. Two specimens: ROM 114CM (3·40 mm long) and NBMG 9879. Possibly conspecific are USNM 483174 and NBMG 9876–9877.

Diagnosis. Species of *Beyrichona* having ventral part of lateral outline strongly acuminate and valve height equal to or greater than valve length.

Description. Weakly postplete, equivalved carapace with triangular lateral outline. Ventral part of lateral outline acuminate. Dorsal margin straight, slightly less than maximum valve length. Trilobate; anterior, central and posterior lobes developed dorsally, ventrally confluent with broadly convex lateral surface, separated by weak, narrow sulci. Anterior lobe extends slightly above the dorsal margin. Central lobe broad and gently convex. Surface smooth.

Remarks. *B. papilio* differs from congeneric species by its markedly acuminate ventral lateral outline. The subdued lobation of the holotype appears to be due to post-mortem flattening of the shell; the lobation of flattened specimens of the co-occuring *B. tinea* resembles that of *B. papilio*.

The USNM suite of four specimens labelled *B. papilio* (*ex*. USNM 50501), from Hanford Brook, New Brunswick, comprises three indeterminate specimens (USNM 483175–483177) and a posterodorsally incomplete right valve (USNM 483174; Ulrich and Bassler 1931*a*, pl. 6, fig. 11; Pl. 4, fig. 10) which preserves a ventrally acuminate lateral outline like that of the holotype, but whose anterodorsal lobation resembles that of *B. tinea*. A posteriorly attenuated valve (NBMG 9876; Pl. 5, fig. 1) might also be referable to *B. papilio*.

The syntypes of *Beyrichona triangula* Matthew, 1895 are a small, dorsally incomplete, indeterminate valve, ROM 26288 B (referred to *B. papilio* by Ulrich and Bassler 1931*a*, p. 44, pl. 7, fig. 6); a now lost specimen (also referred to *B. papilio*; Ulrich and Bassler 1931*a*, p. 44, pl. 7, fig. 7; possibly represented by a poor latex mould associated with ROM 26288 B); and a right valve, ROM 119CM (Ulrich and Bassler 1931*a*, pl. 7, fig. 8; herein designated lectotype, Pl. 5, fig. 5), clearly referable to *B. tinea*.

Occurrence. Type locality and stratum.

EXPLANATION OF PLATE 5

Fig. 1. *Beyrichona papilio* Matthew, 1886?; NBMG 9876; Early Cambrian Hanford Brook Formation, Hanford Brook, New Brunswick; posteriorly and dorsally incomplete internal mould of right valve; ×10.

Figs 2–12. *Beyrichona tinea* Matthew, 1886; Early Cambrian Hanford Brook Formation, Hanford Brook, New Brunswick. 2, USNM 483168; partly exfoliated right valve (Ulrich and Bassler 1931*a*, pl. 6, fig. 9); ×10. 3, ROM 118CM A; partly exfoliated right valve; ×10. 4, ROM 118CM B; posterodorsally incomplete internal mould of left valve, lectotype (?Matthew 1886, pl. 6, fig. 21, 21*a–b* [specimen embedded in wax]); ×9. 5, ROM 119CM; posterodorsally incomplete internal mould of right valve of ?carapace (specimen embedded in rock; lectotype of *Beyrichona triangula* Matthew, 1895), stereo-pair; ×9. 6, NBMG 9815; internal mould of left valve, stereo-pair; ×9. 7–9, NBMG 10023; internal mould of carapace; ×10. 7, dorsal view, stereo-pair. 8, ventral view. 9, right valve. 10, ROM 613CM(B); mostly exfoliated, dorsally damaged right valve (lectotype of *Beyrichona planata* Matthew, 1895, pl. 7, fig. 7); ×8. 11, ROM 118CM D; internal mould of left valve; ×9. 12, NBMG 9820; internal mould of left valve; ×9.

All are lateral views except figs 3, 8.

SIVETER and WILLIAMS, *Beyrichona*

Beyrichona tinea Matthew, 1886

Plate 5, figures 2–12

1886 *Beyrichona tinea*, n. sp., Matthew, p. 66, pl. 6, fig. 21, 21*a–b*.
1894 *Beyrichona tinea*, Matt.; Matthew, p. 97.
1895 *Beyrichona tinea*; Matthew, p. 134, pl. 7, fig. 6*a–c*.
1895 *Beyrichona planata* n. sp., Matthew, p. 134, pl. 7, fig. 7.
*non*1895 *Beyrichona triangula* n. sp., Matthew, p. 135, pl. 7, fig. 5 [= indeterminate composite of more than one specimen].
1895 *Beyrichona ovata* n. sp. Matthew, p. 135, pl. 7, fig. 8.
1895 *Beyrichona rotundata*, n. sp. Matthew, p. 136, pl. 7, fig. 9.
?1898 *Beyrichia*(?) *primaeva*, n. sp., Matthew, p. 133, pl. 1, fig. 2*a–c*.
1902 *Beyrichona ovata*; Matthew, p. 458 [referred to *Escasona* Matthew, 1902].
?1905 *Beyricona gevalensis* n. sp., Wiman, p. 46, pl. 1, figs 29–30, 35.
?1905 *Beyrichona*? *alta* n. sp., Wiman, p. 47, pl. 3, fig. 53.
1905 *Beyrichona linea* [*sic*] Matt.; Wiman, p. 46.
1931*a* *Beyrichona tinea* Matthew; Ulrich and Bassler, p. 43, pl. 6, figs 1–2, 5–7; pl. 7, fig. 4.
1931*a* *Beyrichona tinea triangula* (Matthew); Ulrich and Bassler, p. 44 [*partim*], pl. 6, figs 9–10, pl. 7, fig. 8 [= ROM 119CM]; ?pl. 6, fig. 8 [specimen lost]; *non* pl. 7, fig. 9 [original figure of Matthew 1895, pl. 7, fig. 5 = composite of more than one specimen].
1931*a* *Beyrichona tinea planata* (Matthew); Ulrich and Bassler, p. 45, pl. 6, figs 3–4; pl. 7, figs 1–3.
1931*a* *Beyrichona rotundata* Matthew; Ulrich and Bassler, p. 46, pl. 6, fig. 12; pl. 7, figs 11–12.
1931*a* *Beyrichona ovata* Matthew; Ulrich and Bassler, p. 47, pl. 7, fig. 10.
*non*1934 *Beyrichona tinea* Matthew?; Cobbold, p. 380, pl. 41, fig. 9 [= *Bradoria*? fragment].
1961 *B. tinea*; Sylvester-Bradley, p. Q102, fig. 38,1*b*.
*non*1971 *Beyrichona rotundata* Matthew; Taylor and Rushton, p. 6 [= indeterminate].
1993c *Beyrichona rotundata* Matthew; Hinz-Schallreuter, p. 417 [referred to *Tagra* gen. nov.].

Lectotype. Designated herein; ROM 118CM B (probably original of Matthew 1886, pl. 6, fig. 21, 21*a–b*; Ulrich and Bassler 1931*a*, pl. 7, fig. 4; Pl. 5, fig. 4); part and counterpart of a left valve; Early Cambrian *Protolenus* trilobite Biozone, Hanford Brook Formation, Hanford Brook, New Brunswick.

Material. One hundred and forty-six specimens: ROM 118CM A–E (Matthew's syntypes), 119CM, 120CM(A)–120CM(B), 137CM, 613CM(A) A, C–D, 613CM(B), 26287, 26290. USNM 50500, 483167–483173; NBMG 3299–3302, 3330 (labelled 'type'), 8080–8092, 9814–9868, 9878 and 10023; OUM AT.90–AT.106. NBMG 8093–8097 include specimens of *B. tinea* but have been unavailable to us for study.

Measurements. Hanford Brook Formation specimens are 2·44–5·83 mm long.

Diagnosis. Species of *Beyrichona* with prominent anterior lobe and sub-circular anterodorsal sulcus. Ventral part of lateral outline gently convex. Valve length subequal to or less than valve height.

Description. Subamplete to slightly postplete, equivalved carapace, lacking posterior gape. Lateral outline subtriangular. Commissure straight. Dorsal margin straight, and almost equal to valve length. Hinge-line well developed. Lateral outline has curve anterodorsally. Trilobate; lobes developed dorsally, confluent ventrally with broadly convex lateral surface. Prominent anterior lobe overreaching dorsal margin. Sub-circular anterodorsal sulcus. Central lobe broad, gently convex. Subdued, gently convex posterior lobe. Posterior sulcus narrow, weakly developed. Marginal ridge entire between cardinal corners, broadest anterodorsally. Surface smooth.

Remarks. The carapace of *B. tinea* has a phosphatic component (EDX analysis). *B. tinea* shows variation in length-height ratio, lateral outline and the development of the anterior and posterior lobes and the marginal ridge. However, particular variants do not cluster at specific horizons or localities. *B. tinea* differs from *B. papilio* by its more gently convex ventral lateral outline.

Conspecific with *B. tinea* are *B. planata*, *B. rotundata* and *B. ovata* (all Matthew, 1895). Syntypes of *B. planata* are three left valves (ROM 613CM(A) A, C–D); a right valve, ROM 613CM(B)

(Matthew 1895, pl. 7, fig. 7; Pl. 5, fig. 10; designated lectotype by Ulrich and Bassler 1931*a*, p. 117); and several additional specimens (NBMG 3299). Syntypes of *B. rotundata* are two ventrally incomplete left valves, ROM 120CM(A) (Matthew 1895, pl. 7, fig. 9; designated lectotype by Ulrich and Bassler 1931*a*, p. 117, pl. 7, figs 11–12) and ROM 120CM(B). The type specimen of *B. ovata*, ROM 137CM (Matthew 1895, pl. 7, fig. 8; holotype by monotypy), is a posteriorly incomplete left valve.

Specimens from the type Hanford Brook Formation which Ulrich and Bassler (1931*a*) referred to *B. rotundata* (USNM 50500; incomplete right valve), *B. tinea* (USNM 483169–483170, right valve and carapace, *ex* USNM 50549) and *B. tinea planata* (USNM 483171–483173, two right valves and a left valve, *ex* USNM 50491) are conspecific with *B. tinea*. Of the three USNM specimens (*ex* USNM 50550) referred to *B. tinea triangula* by Ulrich and Bassler two are referable to *B. tinea* (USNM483168, USNM 483167; 1931*a*, pl. 6, figs 9–10; Pl. 5, fig. 2) and one is apparently lost (1931*a*, pl. 6, fig. 8).

In valve outline, size, lobal morphology and the presence of a marginal ridge, the ?Early Cambrian *Beyrichona gevalensis* Wiman, 1905 (PMU B20; 2·78 mm long) from Scandinavia is very similar to juveniles of *B. tinea* and may be conspecific. The Scandinavian *Beyrichona alta* Wiman, 1905 (PMU B97; 2·33 mm long) appears to differ from *B. gevalensis* by lacking a marginal ridge; however, the absence of this feature in specimens of *B. tinea* from the Hanford Brook Formation is often a preservational factor and this may also be the case in *B. alta*.

Occurrence. Known with certainty only from the type locality and stratum. Possibly from the ?Lower Cambrian of Scandinavia (Wiman 1905).

Beyrichona triceps (Matthew, 1903)

Plate 6, figures 1–3

?1881 *Beyrichia hollii*, sp. nov., Jones, p. 343, pl. 10, fig. 7.
1903 *Beyrichia triceps*, n. sp. Matthew, p. 48, 219, pl. 16, fig. 4*a–c*.
1931*a* *Beyrichona triceps* (Matthew); Ulrich and Bassler, p. 46, pl. 6, figs 13–15.
1973 *Beyrichona*-like bradoriid; Bulman and Rushton, pl. 6, fig. 14.

Lectotype. Designated herein, a partly exfoliated right valve, ROM 905CM (?original of Matthew 1903, pl. 16, fig. 4*a–c*; Pl. 6, fig. 1); Mid Cambrian Maclean Brook Formation, Maclean Brook, near Mira River, Cape Breton, Nova Scotia. The syntypes of *B. triceps* are: ROM 57CM, 59CM, 60CM, 905CM; NBMG 3303, 3304.

Material. Some 26 Canadian specimens: ROM 57CM, 59CM, 60CM, 905CM; USNM 50502–50504; NBMG 3303–3304, 3298, 4416, 9190. Six British specimens: BGS DEW9480, GSM 74178, RX 2277–2278, CS.1670; BRSMG CC.1824.

Measurements. Maclean Brook Formation specimens are 1·61–2·5 mm long.

Diagnosis. Small, elongate, markedly trilobate species of *Beyrichona* with prominent finger-like anterior and posterior lobes separated from a broad central lobe by well-developed sulci. Valve length greater than valve height.

Description. Carapace subamplete to postplete, valve length greater than height. Lateral outline sub-triangular. Outline of dorsal margin straight, hinge-line well developed. Anteroventral part of lateral outline almost straight, remainder of lateral outline convex. Trilobate; finger-like prominent anterior and less prominent posterior lobes. Central lobe broad and lower in relief than other lobes. Anterior sulcus deep and narrow, posterior sulcus broad and shallow. Lobes confluent ventrally with convex lateral valve surface. Narrow, entire marginal ridge. Valve surface smooth.

Remarks. *Beyrichona triceps* is smaller, has more prominent lobation and is more elongate than *B. papilio* and *B. tinea*.

The mid Cambrian *B. hollii* Jones, 1881, from the Menevian Beds of South Wales, is known only from a single left valve (NHM I.144) which strongly resembles *B. triceps*. More specimens of *B. hollii* are needed to establish its ontogenetic and intraspecific variation, but it might represent a senior synonym.

Specimens from the basal Ordovician of the Welsh Borderland and English Midlands are conspecific with *B. triceps* (Williams and Siveter in press).

Occurrence. In North America from the Mid Cambrian part (*P. forchhammeri* trilobite Biozone) of the Maclean Brook Formation, near Mira River, Cape Breton and *c*. 2·5 km west of Marion Bridge, Cape Breton, Nova Scotia (Matthew 1903; Ulrich and Bassler 1931*a*). In England from the early Ordovician Tremadoc Series of Gloucestershire, Shropshire and Warwickshire (Williams and Siveter in press). Possibly from the Middle Cambrian of South Wales, UK (Jones 1881). We have found no evidence to support Ulrich and Bassler's (1931*a*, p. 47) report of *B. triceps* from the lower Middle Cambrian at Dugald Brook, Nova Scotia.

Family HIPPONICHARIONIDAE Sylvester-Bradley, 1961

Diagnosis. Small to medium sized (adults *c*. 2–6 mm long), subamplete to slightly postplete Bradoriida with sub-triangular lateral outline, straight dorsal margin and well developed hinge-line. Tri- to multilobate; anterior and posterior lobes may be discrete or ventrally fused to form broad connecting lobe. Trilobate forms have small, sinusoidal central lobe situated adjacent to dorsal margin and anteriorly of valve mid-length. Multilobate forms with up to seven ventrally confluent lobes. Admarginal ridge entire.

Remarks. In lateral valve outline, general lobation and possession of a straight dorsal margin, the hipponicharionids resemble certain palaeocope ostracods; indeed, without evidence from soft parts it is impossible to confirm that they truly are a separate group. Herein we follow the convention of treating hipponicharionids with the Bradoriida. Hipponicharionids differ from palaeocopes by being larger (up to *c*. 6 mm long), exclusively non-dimorphic and by having a phosphatic carapace composition. The detailed structures of the hipponicharionid hinge are not preserved, thus preventing comparison with the hinge structures typical of palaeocopes. Furthermore, adductorial muscle scars are not documented from hipponicharionids. Evolutionary trends within hipponicharionids include confluence of the lobes and a development of multilobate from trilobate forms.

Genus HIPPONICHARION Matthew, 1886

Type species. By monotypy; *Hipponicharion eos* Matthew, 1886, from the Early Cambrian Hanford Brook Formation, Hanford Brook, New Brunswick.

Diagnosis. Medium-sized (adults *c*. 6 mm long), granulose, trilobate hipponicharionids with discrete lobes. Broad, strongly developed, subequally sized anterior and posterior lobes subparallel to the anterior and posterior lateral valve outline. Central lobe has low relief, situated anterior of valve midlength, elongate perpendicular to dorsal valve margin. Anterior sulcus well developed; posterior sulcal depression shallow. Admarginal ridge narrow, continuous between cardinal corners, separated from lobate area by a well developed furrow.

Remarks. Hinz-Schallreuter (1993*b*; see also Siveter *et al*. 1994*b*) used lobal morphology to differentiate hipponicharionid genera. Based on the confluence of the anterior and posterior lobes she assigned the Baltic species *Hipponicharion matthewi* Wiman, 1905 and *Hipponicharion confluens* Ulrich and Bassler, 1931*a* from Canada to her genus *Wimanicharion*. The Baltic genus *Andresia* Hinz-Schallreuter, 1993*b* is distinguished from *Hipponicharion* by having several loculi at the ventral termination of the central lobe. The mid Cambrian, Canadian species *Hipponicharion plicatum* and *Hipponicharion parvum* (both Ulrich and Bassler, 1931*a*) are herein assigned to *Septadella* Stubblefield, 1933.

Occurrence. Lower Cambrian of New Brunswick (Matthew 1886, 1895) and southern Britain (Cobbold 1936; Williams and Siveter in press).

Hipponicharion eos Matthew, 1886

Plate 6, figures 4–8; Plate 7, figures 1–2

1886 *Hipponicharion eos*, n. sp., Matthew, p. 64, pl. 6, fig. 19, 19*a–b*.
1894 *Hipponicharion cavatum*, n. sp., Matthew, p. 99, pl. 17, fig. 3*a–b*.
1894 *Hipponicharion minus*, n. sp., Matthew, p. 99, pl. 17, fig. 4*a–b*.
1895 *Hipponicharion eos*, Matthew, p. 132, pl. 7, fig. 1*a–c*.
1895 *Hipponicharion cavatum*; Matthew, p. 133, pl. 7, fig. 2*a–b*.
1895 *Hipponicharion minus*; Matthew, p. 133, pl. 7, fig. 3*a–b*.
1905 *Hipponicharion eos*; Wiman, p. 49.
1905 *Hipponicharion cavatum*; Wiman, p. 49.
1905 *Hipponicharion minus*; Wiman, p. 49.
1931*a* *Hipponicharion eos* Matthew; Ulrich and Bassler, p. 62, pl. 7, fig. 31.
1931*a* *Hipponicharion cavatum* Matthew; Ulrich and Bassler, p. 63, pl. 7, fig. 32.
1931*a* *Hipponicharion minus* Matthew; Ulrich and Bassler, p. 63, pl. 2, figs 14–16; pl. 7, fig. 30.
1956 *Hipponicharion minus* Matthew; Lochman, p. 1357.
1956 *Hipponicharion cavatum* Matthew; Lochman, p. 1357.
1961 *H. eos*; Sylvester-Bradley, p. Q102, fig. 39.1*a*.
1961 *H. minus*; Sylvester-Bradley, p. Q102, fig. 39.1*b*.
1993*b* *Hipponicharion eos* Matthew, 1886; Hinz-Schallreuter, p. 427, fig. 15. 3.

Lectotype. Designated Hinz-Schallreuter (1993*b*, p. 427, fig. 15); a posteriorly incomplete adult right valve, ROM 267CM (Matthew 1886, pl. 6, fig. 19, 19*a–b*; Ulrich and Bassler 1931*a*, pl. 7, fig. 31; Pl. 6, fig. 4); Early Cambrian Hanford Brook Formation, Hanford Brook, New Brunswick.

Material. Twenty-four specimens: ROM 267CM; 268CM A–C; 795CM A–B; USNM 483178–80 (ex USNM 50552); NBMG 8100–8103, 9883–9892; OUM AT.110-AT.111.

Measurements. Specimens from the type Hanford Brook Formation are 3·33–5·71 mm long; the lectotype is the largest specimen.

Diagnosis. Species of *Hipponicharion* with sinusoidal central lobe and a row of closely spaced granules on the admarginal side of the anterior and posterior lobes.

Description. Subamplete to slighly postplete carapace with sub-triangular lateral outline. Hinge-line almost equal to maximum valve length. Lateral outline evenly convex, without marked curves. Subequally sized, dorsally broadening anterior and posterior lobes, subparallel to adjacent valve outline, separated ventrally by a narrow gap. Smaller, sinusoidal central lobe. Well developed anterior sulcus shallows ventrally. Posterior sulcus broad and shallow. Admarginal ridge narrow, entire, demarcated from lobal area by a deep, narrow groove. Surface granulose. Admarginal side of anterior and posterior lobes each have row of closely spaced granules (Pl. 7, fig. 2).

Remarks. The carapace of *H. eos* has a phosphatic component (EDX analysis). We consider that *Hipponicharion cavatum* and *Hipponicharion minus* (both Matthew, 1894) are synonymous with *H. eos*. None of the three syntypes of *H. minus*, ROM 268CM A–C (see Pl. 6, fig. 8 and Pl. 7, fig. 2), can be identified as Matthew's original figured specimen (1894, pl. 17, fig. 4*a–b*; designated 'type' by Ulrich and Bassler 1931*a*, p. 118, pl. 7, fig. 30), which we presume is lost. *Hipponicharion cavatum* is based on a sheared, part and counterpart carapace, ROM 795CM A–B (Matthew 1894, pl. 17, fig. 3*a–b*; see Pl. 7, fig. 1; herein designated lectotype).

The lectotype of *H. eos* has a series of faint radiating lines originating from the ventral termination of the central lobe (Pl. 6, fig. 4), which may represent the impression of vascular structures (Vannier *et al*. in press); similar features are known from other bradoriids. The granulose ornament, particularly the rows of granules along the anterior and posterior lobes, are observed only in well-preserved specimens. Lobal structures are fully developed in our smallest specimens (*c.* 3·0 mm).

Occurrence. Type locality and stratum.

Genus SEPTADELLA Stubblefield, 1933

Type species. By original designation; *Septadella jackmanae* Stubblefield, 1933, from the early Ordovician Tremadoc Series, Gloucestershire, England.

Diagnosis. Small (adults *c.* 2–3 mm long) hipponicharionids with up to seven, ventrally confluent lobes.

Remarks. In common with other hipponicharionids *Septadella* is subamplete, has a straight dorsal margin, a well-defined hinge-line, a narrow, entire admarginal ridge and prominent dorsoventrally orientated lobation in which the anteriormost and posteriormost lobes are subparallel to the adjacent lateral valve outline.

Occurrence. Middle Cambrian of Nova Scotia (herein) and lower Ordovician of southern Britain (Stubblefield 1933).

Septadella plicatum (Ulrich and Bassler, 1931*a*)

Plate 7, figure 5, ?figure 4

1931*a* *Hipponicharion plicatum*, new species, Ulrich and Bassler, p. 64, pl. 2, fig. 12.
?1931*a* *Hipponicharion parvum*, new species, Ulrich and Bassler, p. 64, pl. 2, fig. 13.
?1933 *Septadella jackmanae* sp. nov., Stubblefield, p. 371, fig. 2.
1984 *Vojbokalina parva* (Ulrich and Bassler, 1931); Melnikova, p. 36.
1984 *Vojbokalina plicata* (Ulrich and Bassler, 1931); Melnikova, p. 36.
1993*b* *Hipponicharion plicatum* Ulrich and Bassler, 1931; Hinz-Schallreuter, p. 420.

Holotype. Abraded, internal mould of right? valve, USNM 81624 (Ulrich and Bassler 1931*a*, p. 64, pl. 2, fig. 12; Pl. 7, fig. 5); Mid Cambrian Maclean Brook Formation, McNeil Brook, east of Marion Bridge, Cape Breton, Nova Scotia.

Material and measurements. The holotype (2·8 mm long) and possibly an additional, right? valve, USNM 50551 (holotype of *Hipponicharion parvum* Ulrich and Bassler, 1931*a*, pl. 2, fig. 13; Pl. 7, fig. 4; 1·5 mm long).

Diagnosis. Species of *Septadella* with seven, subequally developed, ventrally confluent lobes, all of which appear to reach the dorsal margin.

EXPLANATION OF PLATE 6

Figs 1–3. *Beyrichona triceps* (Matthew, 1903); Mid Cambrian Maclean Brook Formation, Maclean Brook, near Mira River, Cape Breton, Nova Scotia. 1, ROM 905CM; partly exfoliated right valve, lectotype (?Matthew 1903, pl. 16, fig. 4*a–c*), stereo-pair; × 20. 2–3, ROM 57CM; two, mostly exfoliated left valves on a single rock slab, stereo-pairs; × 20.

Figs 4–8. *Hipponicharion eos* Matthew, 1886; Early Cambrian Hanford Brook Formation, Hanford Brook, New Brunswick. 4, ROM 267CM; internal mould of posteriorly incomplete right valve, lectotype (Matthew 1886, pl. 6, fig. 19, 19*a–b*), stereo-pair; × 9. 5, NBMG 9884; internal mould of right valve, stereo-pair; × 10. 6, USNM 483179; internal mould of left valve, anterior lobe damaged (Ulrich and Bassler 1931*a*, pl. 2, fig. 15); × 10. 7, USNM 483178; internal mould of left valve, posterior lobe damaged (Ulrich and Bassler 1931*a*, pl. 2, fig. 14); × 10. 8, ROM 268CM A; internal mould of right valve, anterior and posterior lobes damaged (syntype of *Hipponicharion minus* Matthew, 1894); × 10.

All are lateral views.

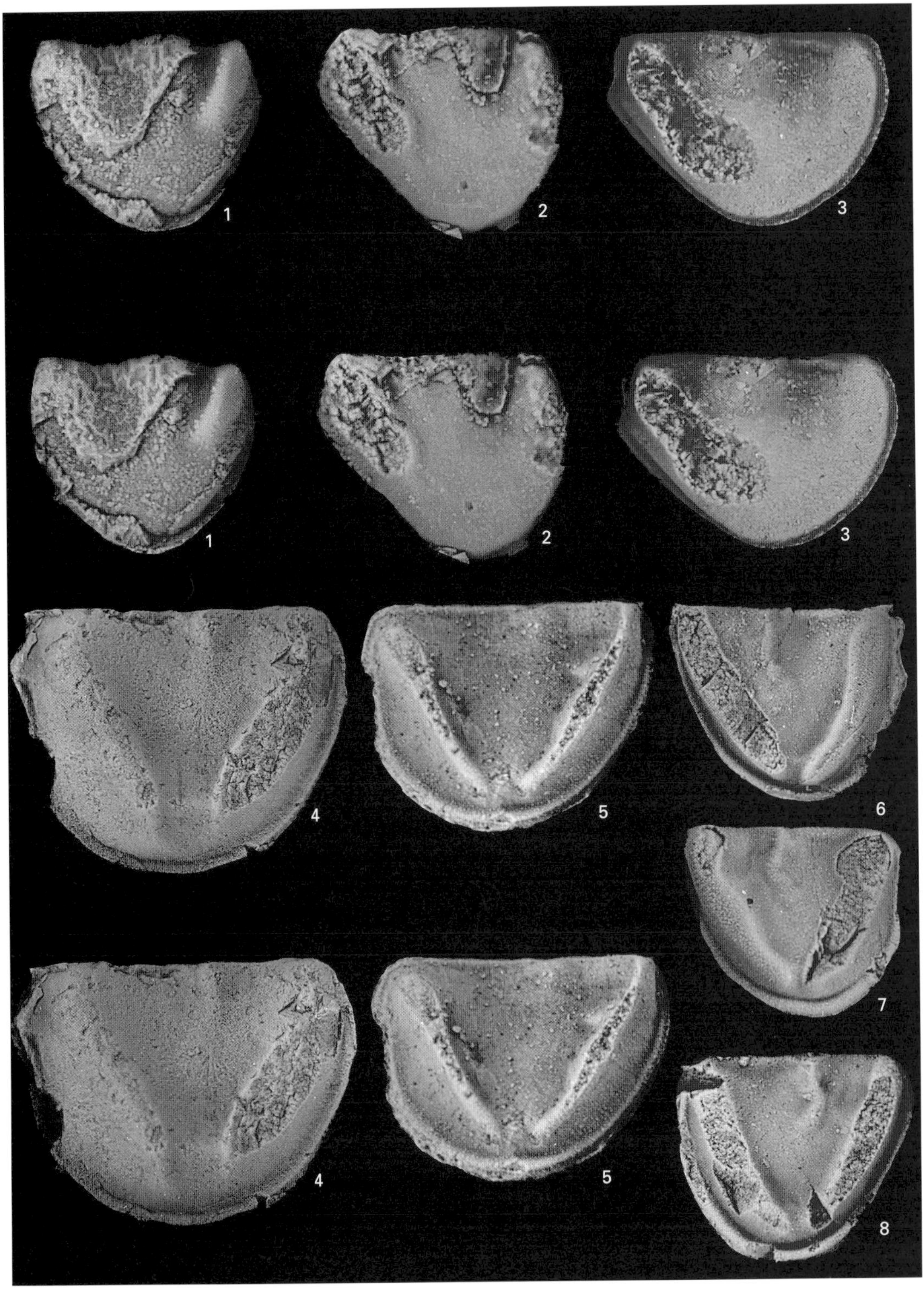

SIVETER and WILLIAMS, *Beyrichona*, *Hipponicharion*

Description. Subamplete, equivalved carapace with sub-triangular lateral outline. Seven, ventrally confluent, narrow, finger-like lobes, all of which reach or overreach the dorsal margin. Central lobe perpendicular to dorsal margin. Anterior and posterior lobes parallel adjacent lateral valve outline. Intervening lobes subperpendicular to dorsal margin. Lobe immediately anterior of posterior lobe most weakly developed.

Remarks. The Mid Cambrian *S. plicatum* has the same lateral valve outline, number of lobes and lobal morphology as *Septadella jackmanae* Stubblefield, 1933 from the lower Ordovician (Tremadoc Series) of southern Britain, and therefore may be synonymous. *S. jackmanae* is known from a much better preserved suite of specimens than *S. plicatum* (Williams and Siveter in press). *Hipponicharion parvum* Ulrich and Bassler, 1931*a*, which appears to have only six lobes, but has a well developed admarginal ridge, is from the same formation and general locality as *S. plicatum* and is probably a juvenile of that species.

Occurrence. Mid Cambrian Maclean Brook Formation, McNeil Brook east of Marion Bridge, and possibly *c*. 2·5 km west of Marion Bridge (*S. parvum*), Cape Breton, Nova Scotia.

Genus WIMANICHARION Hinz-Schallreuter, 1993*b*

Type species. By original designation (Hinz-Schallreuter 1993*b*, p. 422); *Hipponicharion matthewi* Wiman, 1905, from an early Cambrian erratic boulder, Island of Eggegrund, Sweden.

Diagnosis. Small to medium sized (adults *c*. 2–6 mm long), slightly postplete hipponicharionids with confluent anterior and posterior lobes, a discrete central lobe and a narrow, entire admarginal ridge (modified from Siveter *et al*. 1994*b*).

Remarks. *Wimanicharion* differs from the Australian *Cambraechmina* Hinz-Schallreuter 1993*a* and the Chinese *Pseudobeyrichona* Cui, *in* Shu, 1990*a* by having a weak, elongate central lobe, which is oriented perpendicular to the dorsal margin.

EXPLANATION OF PLATE 7

Figs 1–2. *Hipponicharion eos* Matthew, 1886; Early Cambrian Hanford Brook Formation, Hanford Brook, New Brunswick. 1, ROM 795CM B; internal mould of right valve of carapace, left valve skewed behind (lectotype of *Hipponicharion cavatum* Matthew, 1894, pl. 17, fig. 3*a–b*); ×9. 2, ROM 268CM C; internal mould of ventrally incomplete right valve (syntype of *Hipponicharion minus* Matthew, 1894); ×10.

Fig. 3. *Wimanicharion*? *confluens* (Ulrich and Bassler, 1931*a*); USNM 81625; Mid Cambrian Maclean Brook Formation, *c*. 2·5 km west of Marion Bridge, Cape Breton, Nova Scotia; internal mould of dorsally incomplete ?right valve, holotype (Ulrich and Bassler 1931*a*, pl. 2, figs 10–11); ×20.

Fig. 4. *Septadella plicatum* (Ulrich and Bassler, 1931*a*)?; USNM 50551; Mid Cambrian Maclean Brook Formation, *c*. 2·5 km west of Marion Bridge, Cape Breton, Nova Scotia; abraded, internal mould of ?right valve (holotype of *Hipponicharion parvum* Ulrich and Bassler, 1931*a*, pl. 2, fig. 13); ×20.

Fig. 5. *Septadella plicatum* (Ulrich and Bassler, 1931*a*); USNM 81624; Mid Cambrian Maclean Brook Formation, McNeil Brook, east of Marion Bridge, Cape Breton, Nova Scotia; abraded, internal mould of ?right valve, holotype (Ulrich and Bassler 1931*a*, pl. 2, fig. 12); ×15.

Figs 6–9. *Anabarochilina* cf. *australis* (Hinz-Schallreuter, 1993*b*); Mid Cambrian Marjum Formation (R. A. Robison locality 380; figs 6, 8–9) and Weeks Formation (K. R. Evans locality 824; fig. 7), Millard County, Utah. 6, KUMIP 1129849; internal mould of right valve; ×8. 7, KUMIP 1129844; open carapace, stereo-pair; ×8. 8, KUMIP 1129847; internal mould of left valve; ×8. 9, KUMIP 1129848; internal mould of right valve; ×10.

Figs 10–11. *Anabarochilina rotundata* (Walcott, 1887); early Cambrian limestones and interbedded shales, *c*. 3 km south-east of North Greenwich, Washington County, New York State. USNM 17446; carapace, lectotype (Walcott 1887, pl. 1, fig. 9); ×8. 10, silicon rubber cast of external mould of right valve. 11, internal mould of right valve, stereo-pair.

All are lateral views.

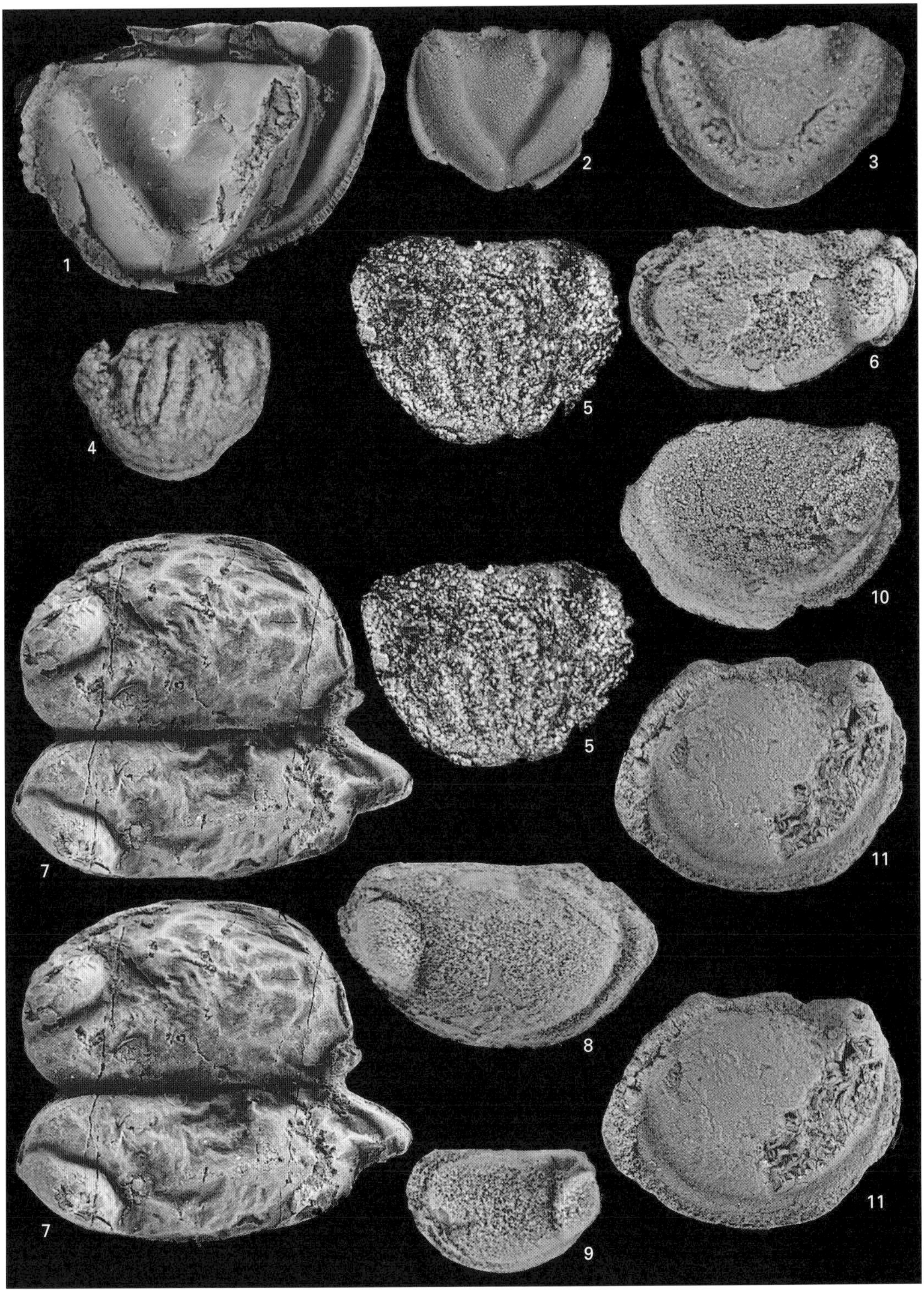

SIVETER and WILLIAMS, bradoriids

Occurrence. Lower Cambrian of Scandinavia and possibly from the Middle Cambrian of Nova Scotia (Siveter *et al*. 1994*b*).

Wimanicharion? *confluens* (Ulrich and Bassler, 1931*a*)

Plate 7, figure 3

1931*a* *Hipponicharion confluens*, new species, Ulrich and Bassler, p. 64, pl. 2, figs 10–11.
1993*b* *Hipponicharion confluens* Ulrich and Bassler, 1931; Hinz-Schallreuter, p. 422 [referred to *Wimanicharion*].
1994*b* *Wimanicharion*? *confluens* (Ulrich and Bassler, 1931); Siveter *et al*., p. 19, text-fig. 1*c*.

Holotype. A dorsally incomplete ?right valve, USNM 81625 (Ulrich and Bassler 1931*a*, p. 64, pl. 2, figs 10–11; Pl. 7, fig. 3); Mid Cambrian Maclean Brook Formation, *c*. 2·5 km west of Marion Bridge, Cape Breton, Nova Scotia.

Material and measurements. Only the holotype: 2·16 mm long.

Remarks. This taxon has confluent anterior and posterior lobes but, as noted by Siveter *et al*. (1994*b*, p. 19), the holotype is damaged dorsally and the presence of a central lobe cannot be verified. Thus, the species is referred to *Wimanicharion* with doubt.

Occurrence. Type locality and stratum.

Family SVEALUTIDAE Öpik, 1968

1968 Svealutidae nov. Öpik, p. 31 [*partim*], *non Alutella* Kobayashi and Kato, 1951 [= Oepikalutidae Jones and McKenzie, 1980], *Carnarvonia* Walcott, 1912.
1980 Svealutidae Öpik, 1968; Jones and McKenzie, p. 207.
1987 Svealutidae Öpik, 1968; Zhang, p. 14.
1993*b* Svealutidae Öpik, 1961 [*sic*]; Hinz-Schallreuter, p. 431.

Diagnosis. Small to large (adults *c*. 2–10 mm long), elongate, subamplete to postplete Bradoriida. Dorsal margin straight or gently curved anterodorsally, hinge-line well developed. Admarginal ridge entire, often widest mid-posteriorly, demarcated from lateral surface by a furrow. Up to three, mid-anterior to anterodorsally situated nodes. Valves smooth and evenly convex.

Remarks. *Anabarochilina* Abushik, 1960 and *Ovaluta* Zhang, 1987 are typical svealutids. Several taxa which have been referred to the Alutidae Huo, 1956 (e.g. see Huo 1956; Huo and Shu 1985; Huo *et al*. 1991), a taxon herein restricted to the type species of *Aluta* Matthew, 1896 (see below), may also belong to the Svealutidae; for example, especially *Tsunyiella* Zhang, 1974 from China and several Chinese species referred to *Aluta* (Walcott 1913).

Genus ANABAROCHILINA Abushik, 1960

1960 *Anabarochilina* gen. nov., Abushik, p. 96.
1961 *Svealuta* gen. nov., Öpik, p. 174.
1993*b* *Eremos* Moberg and Segerberg, 1906; Hinz-Schallreuter, p. 431 [*partim*], *non Eremos bryograptorum* Moberg and Segerberg [= *Eremos* Moberg and Segerberg, 1906].
1993 *Anabarochilina* Abushik, 1960; Siveter *et al*., p. 71.

Type species. By original designation; *Anabarochilina ventriangulosa* Abushik, 1960, from the upper Middle Cambrian of Siberia. We regard *Leperditia primordialis* Linnarsson, 1869 as a senior subjective synonym.

Diagnosis. Medium to large (adults *c*. 5–10 mm long), subamplete to postplete elongate bradoriids. Valves smooth and convex. Hinge-line well developed. Admarginal ridge widest mid-posteriorly, demarcated from lateral surface by a furrow. Up to three nodes, situated in mid-anterior to anterodorsal region.

Remarks. Hinz-Schallreuter (1993*b*) considered *Anabarochilina* to be a junior synonym of *Eremos* Moberg and Segerberg, 1906. We prefer to restrict *Eremos* to its type species, the early Ordovician *Eremos bryograptorum*, which is known only from a single poorly preserved and incomplete specimen (Hinz-Schallreuter 1993*b*, fig. 18.2–18.3).

In size and overall morphology *Anabarochilina* resembles some leperditicope ostracodes (e.g. see Berdan 1984; Copeland 1986). Exfoliated *Anabarochilina* specimens sometimes reveal a network of fine anastomosing lines emanating from the area of anterior lobation. Lines of similar appearance occur in leperditicopes and other bradoriids, where they originate from the adductor muscle scar region and have been interpreted as possible impressions of a vascular system (Vannier *et al.* in press).

An open carapace of *Anabarochilina* from Utah shows a well developed hinge-line (see Pl. 7, fig. 7; cf. Jones and McKenzie 1980).

Occurrence. Lower Cambrian of New York State and upper Middle Cambrian of Utah, USA. Upper Middle Cambrian of southern Britain, Scandinavia, Australia and Kazakhstan, and Middle to Upper Cambrian of Siberia (Jones and McKenzie 1980; Siveter *et al.* 1993; Melnikova *et al.* in press).

Anabarochilina rotundata (Walcott, 1887)

Plate 7, figures 10–11

1887 *Aristozoe rotundata*, n. sp., Walcott, p. 193, pl. 1, fig. 9.
1890 *Aristozoe rotundata* Walcott; Walcott, p. 627, pl. 80, fig. 3.
1929 "*Aristozoe*" *rotundata*.; Gürich, p. 35, text-fig. 1.7 [on p. 37 referred to as *Aristozoe rotundata* Walcott].
1931*a* *Aluta rotundata* (Walcott); Ulrich and Bassler, p. 60, pl. 8, figs 18–19.
1931 *Aristozoe rotundata* Walc.; Kummerow, pp. 244, 253, fig. 14.

Lectotype. Designated herein; a carapace, USNM 17446 (Walcott 1877, pl. 1, fig. 9; Pl. 7, figs 10–11). Although Ulrich and Bassler (1931*a*, p. 60) referred to this specimen as the 'holotype', Walcott (1887, p. 193) apparently had more than one specimen. Early Cambrian limestones interbedded with shales (*Elliptocephala asaphoides* fauna; *Callavia–Protolenus* trilobite Biozones), *c.* 3 km south-east of North Greenwich, Washington County, New York State.

Material and measurements. Only the lectotype: 4·76 mm long.

Diagnosis. Species of *Anabarochilina* with a single anterodorsal node which overreaches the anterior margin.

Description. Weakly postplete. Lateral outline of free margin evenly convex. Lateral surface strongly and evenly convex. Admarginal ridge moderately wide and probably entire, separated from lateral surface by a furrow. Anterodorsal node overreaches anterior margin. Low, narrow, irregular ridges originate from the anterior lobal area. Surface smooth.

Remarks. A. rotundata is the earliest known species of *Anabarochilina*. So-called 'vascular' markings, originating from the anterior lobal area, were noted by Walcott (1887, pl. 1, fig. 9).

Occurrence. Type locality and stratum.

Anabarochilina cf. *australis* (Hinz-Schallreuter, 1993*b*)

Plate 7, figures 6–9

Material. Seven specimens. KUMIP nos 1129844–50, collected from R. A. Robison's localities 380, 732 and 748 and K. R. Evans' locality 824. Locality 380: near the top of the type Marjum Formation (see Hintze and Robison 1975) and at top of cliffs on promontory south of Marjum pass (NW1/4 NW1/4 SW1/4 sec. 18 T.18 S. R. 13 W., Notch Peak 15-minute quadrangle map, U.S. Geological Survey 1960). Locality 732: topmost bed of the Marjum Formation at top of cliff and about 4 metres west of bend in unmetalled road, close to locality 380. Locality 748: 0·6 m above the base of the Weeks Formation on low north-north-west trending ridge approximately 6·5 km east of Marjum Pass and *c*. 400 m north of former US Highway 6–50 (SW1/4 NE1/4 SW1/4 sec. 11 T. 18 S. R. 13 W., Marjum Pass 7·5-minute quadrangle map, US Geological Survey 1972). Locality 824: *c*. 200 m above the base of the Weeks Formation at its type locality (see Hintze and Robison 1975) in North Canyon (formerly Weeks Canyon) and adjacent small quarries (SE1/4 NW1/4 sec. 29 T. 18 S., R. 13 W., Notch Peak 15-minute quadrangle map, US Geological Survey 1960).

Measurements. Specimens from the Weeks and Marjum formations are 3·0–6·5 mm long.

Description. Elongate, postplete carapace. Well developed anterior node overreaching anterior margin; small, subdued anterodorsal node. Admarginal ridge well developed, broadest mid-posteriorly, confluent with furrow demarcating posterior margin of anterior node. Lateral surface convex.

Remarks. Hinz-Schallreuter's (1993*b*) specimens of *A. australis* have a more dorsal anterior node and a better developed anterodorsal node, but may all be juveniles (largest = 5·28 mm long), making direct comparison with the Utah material difficult. A species of *Anabarochilina* from the upper Middle Cambrian of Kazakhstan also resembles the Utah specimens (see Melnikova *et al.* in press).

Occurrence. Late Mid Cambrian *Lejopyge* trilobite Subzone, Weeks and Marjum formations, House Range, Millard County, Utah.

Genus BULLALUTA Copeland, 1986

Type species. By original designation; *Bullaluta kindlei* Copeland, 1986, from the Upper Cambrian of western Newfoundland.

Diagnosis. Small (*c*. 2 mm long), postplete svealutid with a single sub-circular anterior node and an entire admarginal ridge of even width.

Remarks. *Bullaluta* differs from *Anabarochilina* by being much smaller, more postplete and by having a single anterior node and an admarginal ridge of even width.

Copeland (1986) reported a calcium carbonate composition for the carapace of *Bullaluta*. However, his three specimens appear to be mostly internal moulds and the calcium carbonate might therefore be derived from the host sediment. EDX analysis demonstrates a calcium phosphate composition for the carapace of other svealutids (e.g. Scandinavian *Anabarochilina*; see Williams and Siveter in press).

Bullaluta kindlei Copeland, 1986

1986 *Bullaluta kindlei* n. sp., Copeland, p. 402, pl. 42.1, figs 1–6.

Holotype. A mostly exfoliated left valve, GSC 80998 (Copeland 1986, pl. 42.1, figs 3–4), from the Late Cambrian *Cedaria-Crepicephalus* Zone 5 conglomerate, north-east shore of the Cow Head Peninsula, western Newfoundland.

Material and measurements. Three specimens, GSC 80996–80998: 1·7–2·00 mm long.

Diagnosis. As for genus (monotypic).

Description. Elongate, postplete. Well developed sub-circular anterior node and entire admarginal ridge. Lateral surface evenly convex.

Remarks. Well preserved Scandinavian *Anabarochilina* material shows a series of apparently blind pores distributed over the inner surface of the shell. Thus, where the shell is exfoliated, the internal mould appears to show pustulose 'ornament' (e.g. see Siveter *et al.* 1993, pl. 20: 72, fig. 1*a*). Similar features are apparent in *B. kindlei* (e.g. Copeland 1986, pl. 42.1, figs 1–6) and may be a consistent feature of svealutids.

Occurrence. Type locality and stratum.

Genus LIANGSHANELLA Huo, 1956

Type species. By original designation; *Liangshanella liangshanensis* Huo, 1956, from the Lower Cambrian of southern Shaanxi, China.

Diagnosis. Small to medium-sized subovate bradoriids with equivalved, postplete carapace lacking posterior gape. Dorsal margin straight, hinge-line well developed. Admarginal ridge entire and separated from lateral valve surface by a distinct furrow. Anterodorsal node well developed to weak. Centrally situated, sub-circular muscle scar of numerous small pits.

Occurrence. In North America *Liangshanella* occurs in the Middle Cambrian of British Columbia, Alberta (Mount Cap Formation; N. Butterfield, pers. comm.) and Utah. Also occurs in the Lower Cambrian of China (Huo 1956; Huo *et al.* 1991) and possibly the former Soviet Union (cf. *Bradoria*? *sayutinae* Melnikova, 1988; see Melnikova *et al.* in press).

Liangshanella burgessensis sp. nov.

Plate 8, figures 3–5, ?figure 6

?1931*a* *Aluta*(?) *rimulata*, new species, Ulrich and Bassler, p. 61, pl. 5, fig. 16.

Derivation of name. From the Burgess Shale.

Holotype. A juvenile carapace, USNM 272083 (Pl. 8, fig. 3); Mid Cambrian Phyllopod Bed, Burgess Shale, Stephen Formation, Walcott Quarry, British Columbia.
Material. Approximately 350 specimens on rock slabs. ROM 49937–49952; USNM 271995–272021, 272059–272061, 272078–272092. Possibly KUMIP 1129852–1129854.

Measurements. ROM specimens (mostly Walcott and Raymond Quarry talus) are 0·66–4·00 mm long. USNM specimens (from the Phyllopod Bed) are 1·67–4·25 mm long.

Diagnosis. Species of *Liangshanella* with weakly developed, sub-circular anterodorsal node and a sub-circular adductorial muscle scar composed of numerous small pits.

Description. Postplete, equivalved, bivalved carapace. Subovate lateral outline. Hinge-line well developed and straight. Free margin part of lateral outline evenly convex, without marked curves. Admarginal ridge entire, of equal width between cardinal corners and demarcated from lateral surface by a distinct furrow. Centrally situated, sub-circular, adductorial muscle scar composed of numerous small pits. Weak sub-circular anterodorsal node (undeveloped in small juveniles). Lateral surface convex, smooth.

Remarks. The holotype has a distinct sub-central muscle scar corresponding in position, shape and size to adductor muscle scars described from post-Cambrian ostracods such as the leperditicopids (see Berdan 1984).

L. burgessensis occurs in silty mudstones, mostly as open, articulated carapaces lying parallel to lamination, thus suggesting relatively undisturbed and/or rapid burial. Clusters of specimens sometimes occur with small fragments of other organisms (e.g. ROM 49951). In such instances orientation of valves and carapaces can be random, perpendicular to lamination, and possibly rapidly deposited. The ROM and USNM slabs containing *L. burgessensis* are generally too small to include other identifiable Burgess Shale organisms.

L. burgessensis is preserved mostly as moulds or dark grey to black thin films possibly of original shell material. Valves are crumpled, but not pervasively cracked, suggesting that the shell was thin and flexible. Shrivelled specimens may be exuvia. Other specimens, preserved with relief, indicate that valves were originally gently convex overall. EDX analysis indicates that phosphate is a component of the shell but not the host sediment. In some instances pyrite coats the inner surfaces of valves and carapaces.

Like some other Burgess Shale arthropods such as *Isoxys* (Conway Morris 1986; Williams *et al.* 1996), *L. burgessensis* lacks preserved soft tissues; their absence might indicate that most specimens are exuvia. In the early Cambrian Chengjiang *Lagerstätte* of South China, where bradoriids account for 80 per cent. of the entire fauna (Hou and Bergström 1991), rare preservation of bradoriid soft-parts occurs (Hou *et al.* 1996).

In Walcott's (USNM) collections this species is labelled '*Aluta burgessensis*' but this manuscript name was never published. The type and only specimen of *Aluta*(?) *rimulata* Ulrich and Bassler, 1931*a* (USNM 81735), from the Middle Cambrian of Alberta, is lost, but the species appears to resemble closely *L. burgessensis*. Rare, poorly preserved *Liangshanella* from transition beds between the Wheeler Formation and overlying Pierson Cove Formation (*Ptychagnostus atavus* trilobite Subzone) of the southern Drum Mountains, Millard County, Utah (R. A. Robison collection), might also be conspecific with *L. burgessensis* (see Pl. 8, fig. 6).

Occurrence. Mid Cambrian Burgess Shale (*Ptychagnostus praecurrens* trilobite Subzone), Stephen Formation, Raymond and Walcott quarries, near Field, British Columbia. Possibly from the Middle Cambrian of Utah (*Ptychagnostus atavus* trilobite Subzone).

EXPLANATION OF PLATE 8

Fig. 1. *Matthoria troyensis* (Ford, 1873); USNM 17447; Early Cambrian limestones and interbedded shales, *c.* 1 km west of North Greenwich, Washington County, New York State; internal mould of left valve (Walcott 1890, pl. 80, fig. 2*a*), stereo-pair; ×8.

Fig. 2. *Aluta flexilis* Matthew, 1896; ROM 482CM; Mid Cambrian Fossil Brook Member, Chamberlains Brook Formation, Porter Road, New Brunswick; holotype, ?bradoriid valve (Matthew 1896, pl. 15, fig. 4), stereo-pair; ×11.

Figs 3–5. *Liangshanella burgessensis* sp. nov.; Mid Cambrian Burgess Shale, Stephen Formation, above field on west side of ridge between Mount Field and Wapta Peak, *c.* 1·5 km north-east of Burgess Pass (figs 3–4) and talus from below the Walcott and Raymond quarries (fig. 5), British Columbia. 3, USNM 272083; open carapace, holotype, showing possible adductor muscle scar, stereo-pair; ×15. 4, USNM 272080; distorted, open carapace; ×15. 5, ROM 49941; external mould of incomplete, open carapace; ×15.

Fig. 6. *Liangshanella burgessensis*?; KUMIP 1129852; Mid Cambrian transition beds between Wheeler Formation and Pierson Cove Formation, NW1/4 SW1/4 NE1/4 of section 20, T 15 S R 10 W, Drum Mountains Well 7·5-minute quadrangle map (US Geol. Survey 1971), southern Drum Mountains, Millard County, Utah; right valve of flattened carapace; ×30.

Fig. 7. *Vestrogothia*? *trituberculata* (Lochman and Hu, 1960); USNM 138145; Late Cambrian Du Noir Limestone, Sheep Mountain section, Northwest Wind River Mountains, north-eastern Sublette County, Wyoming; internal mould of left valve, holotype (Lochman and Hu 1960, pl. 98, fig. 56); ×45.

Figs 8–9. *Cyclotron lapworthi* (Groom, 1902); Late Cambrian, Maclean Brook Formation, Manuels River, eastern Newfoundland. 8, USNM 56492; internal mould of posteriorly incomplete left valve (holotype of *Polyphyma marginata* Ulrich and Bassler, 1931*a*, pl. 8, figs 28–29); ×12. 9, USNM 481618, internal mould of tectonically elongated right valve (Williams *et al.* 1994*c*, fig. 6*b*); ×12.

All are lateral views.

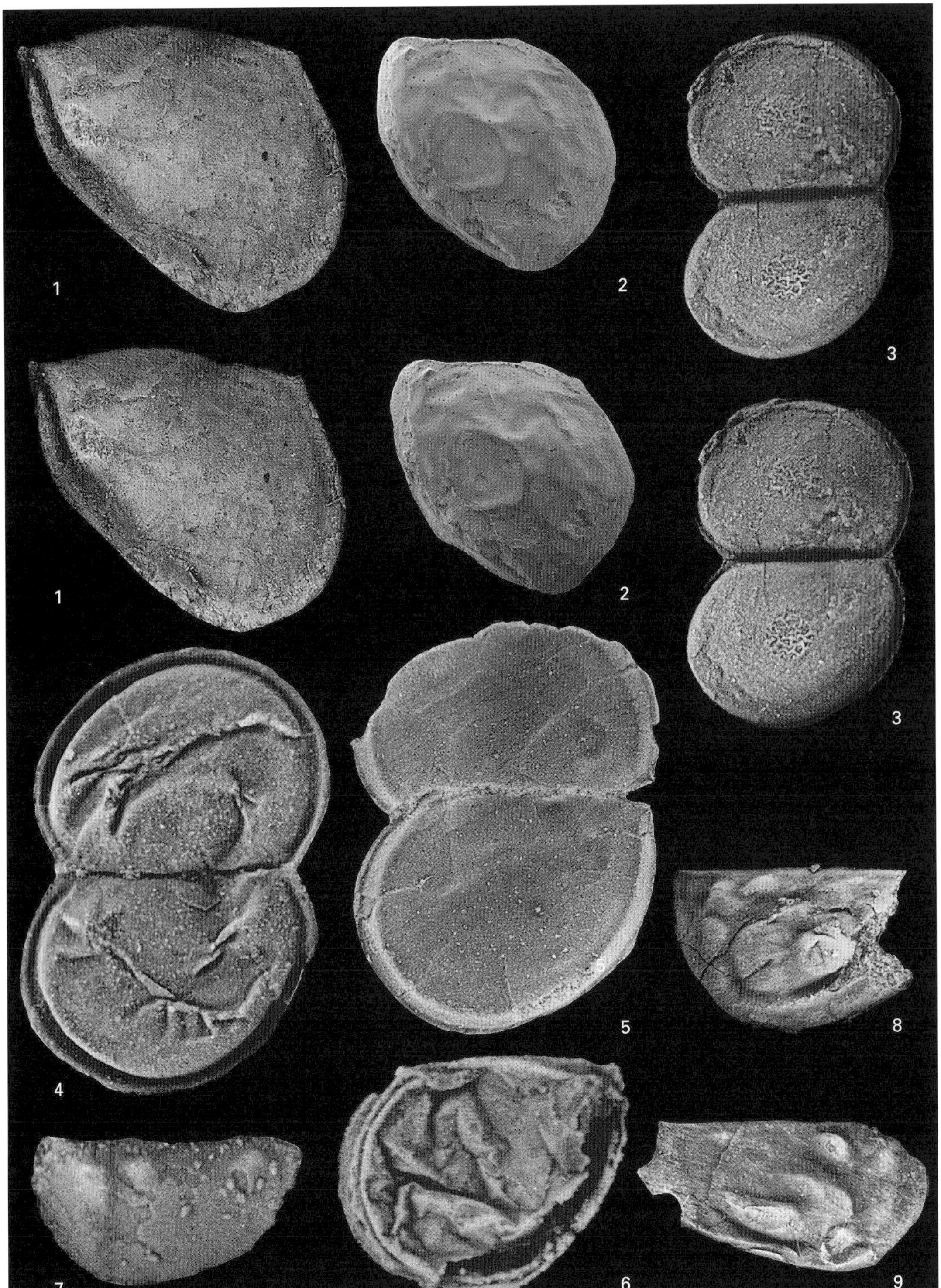

SIVETER and WILLIAMS, bradoriids, *Aluta* and phosphatocopids

Genus MATTHORIA gen. nov.

Derivation of name. Elision of *Matthaeus*, Matthew, and *Bradoria*; in honour of G. F. Matthew.

Type species. Leperditia troyensis Ford, 1873, from the Lower Cambrian of New York State.

Diagnosis. Postplete bradoriids with a cusp on anterior part of dorsal margin, a broad anterodorsal node and an entire admarginal ridge.

Remarks. The postplete shape, concentration of lobal features anterodorsally and development of an admarginal ridge suggest a svealutid affinity. However, the anterodorsal cusp is also reminiscent of Cambriidae (see Williams *et al.* 1994*a*; Siveter *et al.* 1994*a*, 1996).

Congeneric early Cambrian bradoriids include a species from the Purley Shale Formation (*Protolenus* trilobite Biozone) in southern Britain (see Rushton 1966; Williams and Siveter in press) and possibly '*Bradoria*' *ordinata* Melnikova, 1983 from the Lower Cambrian of Siberia (see Melnikova *et al.* in press).

Occurrence. Lower Cambrian of New York State (herein) and southern Britain (Williams and Siveter in press), and possibly from the Lower Cambrian of Siberia (see Melnikova *et al.* in press).

Matthoria troyensis (Ford, 1873)

Plate 8, figure 1

1873 *Leperditia Troyensis*, sp. nov., Ford, p. 138.
1886 *Leperditia troyensis* Ford; Walcott, p. 146, pl. 16, fig. 5.
1887 *Aristozoe troyensis*; Walcott, p. 193, pl. 1, fig. 8.
1890 *Aristozoe troyensis*; Walcott, p. 628, pl. 80, fig. 2, 2*a*.
1929 '*Aristozoe*' *Troyensis*; Gürich, p. 35, text-fig. 1.6 [on p. 37 referred to as *Aristozoe Troyensis* Ford].
1931*a* *Aluta troyensis* (Ford); Ulrich and Bassler, p. 58, pl. 8, figs 15–17.
1961 *A. troyensis*; Sylvester-Bradley, p. Q102, fig. 38.4*b*.
1985 *Aluta troyensis* (Ford); Huo and Shu, p. 62.
1987 *Leperditia troyensis* Ford; Zhang, p. 14.
1991 *Aluta troyensis* (Ford); Huo *et al.* p. 67.

Type specimen. The right valve on which Ford based this species (1873; Walcott 1886, 1887, 1890; Ulrich and Bassler 1931*a*, pl. 8, fig. 15) has not been located in USNM collections; early Cambrian limestones interbedded with shales (*Elliptocephala asaphoides* fauna; *Callavia–Protolenus* trilobite biozones), ridge east of the City of Troy, New York State, USA.

We figure a left valve, USNM 17447, of Walcott 1890 (pl. 80, fig. 2a; Ulrich and Bassler 1931*a*, pl. 8, fig. 17; Pl. 8, fig. 1) from *c.* 1 km west of North Greenwich, Washington County, New York State.

Material and measurements. One specimen, USNM 17447: 6·00 mm long.

Diagnosis. As for the genus.

Description. Postplete carapace with subovate lateral outline. Broad anterodorsal node. Anterior part of dorsal margin is projected dorsally into a broad gently curved cusp. Lateral outline has mid-anterior curve directly below anterior node. Lateral valve surface evenly and gently convex. Admarginal ridge entire, weakly developed ventrally, narrowing towards cardinal corners.

Remarks. A congeneric bradoriid from the early Cambrian part of the Purley Shale Formation of Warwickshire, England (see Rushton 1966; Williams and Siveter in press), differs by having a more amplete elongate shape.

Occurrence. Early Cambrian limestones interbedded with shales (*Elliptocephala asaphoides* fauna; *Callavia–Protolenus* trilobite biozones), at the type locality and *c.* 1 km west of North Greenwich, Washington County, New York State.

Family Uncertain

Genus MONONOTELLA Ulrich and Bassler, 1931*a*

Type species. Subsequently designated Ulrich and Bassler (1931*a*, p. 91); *Primitia*(?) *fusiformis* Matthew, 1895, from the Lower Cambrian of New Brunswick.

Diagnosis. Medium-sized to large (adults *c.* 10 mm long), markedly elongate and postplete bradoriids with a univalved carapace. No lobation, nodation or marginal structures; valves gently and evenly convex. Valves punctate or smooth.

Remarks. In valve outline *Mononotella* resembles *Indiana* but differs in being more postplete and by apparently having a univalved carapace. Other relatively small Cambrian bradoriids/phosphatocopids with a univalved carapace include the early Cambrian Chinese *Dabashanella* Huo, Shu and Fu (*in* Huo *et al.*, 1983) and the mid Cambrian *Pejonesia* Hinz, 1992 from Australia. *Dabashanella* differs from *Mononotella* by its amplete lateral shape and well-developed cardinal spines and is generally much smaller (adults *c.* 1 mm). *Pejonesia* differs from *Mononotella* by having a sharp bend where the two halves of the carapace meet dorsally and by having a latero-admarginal ridge. Other Cambrian arthropods with a univalved carapace or partially developed dorsal hinge are much larger (e.g. *Combinivalvula* Hou, 1987).

Occurrence. Lower Cambrian of New Brunswick (Ulrich and Bassler 1931*a*). Other occurrences (e.g. Shu 1990*a*) need verification.

Mononotella fusiformis (Matthew, 1895)

Plate 4, figures 1–3

1895 *Primitia*(?) *fusiformis* Matthew, p. 137, pl. 8, fig. 3*a–b*.
1902 *Primitia*(?) *fusiformis*; Matthew, p. 461 [referred to *Indiana*].
1931*a* *Mononotella fusiformis* (Matthew); Ulrich and Bassler, pp. 92–93 [*partim*], pl. 8, figs 22–25; *non* specimen from Dugald Brook [= indeterminate].
1931*a* *Entomidella*(?) *buprestis* (Salter); Ulrich and Bassler, p. 98 [*partim*], pl. 8, fig. 21, *non* pl. 8, fig. 20 [= *Entomidella buprestis* (Salter)].
1961 *M. fusiformis*; Sylvester-Bradley, p. Q103, fig. 39.4*a–b*.

Holotype. A right half of the carapace, ROM 132CM (type by monotypy; Matthew 1895, pl. 8, fig. 3*a–b*; Pl. 2, figs 1–2); Early Cambrian *Protolenus* trilobite Biozone, Hanford Brook Formation, Hanford Brook, New Brunswick.

Diagnosis. As for the genus.

Material and measurements. Seven mostly incomplete 'valves' and carapaces. ROM 132CM; USNM 56499, 56501, 487782. The holotype is 7 mm long.

Description. Markedly elongate, postplete lateral shape. Dorsal margin straight, approximately half carapace length. Posterodorsal and anteroventral part of lateral outline straight to slightly convex. Carapace appears to be univalved. Shell thin. Valves broadly convex, non-lobate and finely punctate.

Remarks. The holotype, a right half of a carapace, has a straight dorsal margin (cf. Ulrich and Bassler 1931*a*, p. 92). Ulrich and Bassler's specimens from Hanford Brook, New Brunswick, are incomplete or deformed, and suggest that the carapace was poorly mineralized and/or flexible. Carapaces (e.g. USNM 56499) appear to show no discrete dorsal hinge, thus indicating that *Mononotella* was univalved.

Conspecific with *M. fusiformis* is the Hanford Brook specimen (USNM 56501) which Ulrich and Bassler (1931*a*) referred to *Entomidella*(?) *buprestis* (Salter). Ulrich and Bassler's (1931*a*, p. 93; USNM 56500) incomplete 'valve' from the Dugald Formation (division E3b), which they tenatively referred to *M. fusiformis*, is indeterminate.

Occurrence. Type locality and stratum.

Taxa of Uncertain Status

Genus ALUTA Matthew, 1896

1896 *Aluta*, n. gen., Matthew, p. 198.
1931*a* *Aluta* Matthew; Ulrich and Bassler, p. 49 [*partim*; only *Aluta flexilis* Matthew].
1961 *Aluta* Matthew, 1896; Sylvester-Bradley, p. 102 [*partim*], *non A. troyensis* (Ford) [= *Matthoria* gen. nov.].
1968 *Aluta* Matthew, 1896; Öpik, p. 26.
1985 *Aluta* Matthew; Huo and Shu, p. 61 [*partim*; only *Aluta flexilis* Matthew].
1987 *Aluta*; Zhang, p. 14.
*non*1991 *Aluta* (Matthew) [*sic*]; Huo *et al*., p. 67.
1993*b* *Aluta* Matthew, 1896; Hinz-Schallreuter, p. 431.

Type species. By monotypy; *Aluta flexilis* Matthew, 1896, from the Mid Cambrian Fossil Brook Member, Chamberlains Brook Formation, New Brunswick.

Remarks. Several species world-wide have been assigned to *Aluta*. Öpik (1968) restricted the family Alutidae Huo, 1956 to the holotype of the type species of *Aluta*. We consider that the latter specimen is not identifiable as a bradoriid, and regard *Aluta* as a *nomen dubium*.

Aluta flexilis Matthew, 1896

Plate 8, figure 2

1896 *Aluta flexilis*, n. sp., Matthew, p. 198, pl. 15, fig. 4.
1931*a* *Aluta flexilis* Matthew; Ulrich and Bassler, p. 52, pl. 8, figs 13–14.
1968 *Aluta flexilis* Matthew; Öpik, p. 26.
1985 *Aluta flexilis* Matthew; Huo and Shu, p. 61.
1987 *Aluta flexilis*; Zhang, p. 14.

Holotype. Specimen ROM 482CM (Matthew 1896, pl. 15, fig. 4; Pl. 8, fig. 2); Mid Cambrian Fossil Brook Member, Chamberlains Brook Formation, Porter Road, New Brunswick.

Material and measurements. Only the holotype (3·1 mm long).

Remarks. The only specimen of *A. flexilis* is distorted, has no distinctively identifiable dorsal margin and has possible puncta on the lateral surface. Its identification as a bradoriid valve and its possible orientation cannot be confirmed; indeed, it may be a trilobite fragment, which are abundant at the type locality.

Ulrich and Bassler (1931*a*) considered Matthew's (1896) original figure of *A. flexilis* to be upside down, which they reoriented to show a postplete outline and straight dorsal hinge and to note its similarity to some Chinese bradoriids (see also Walcott 1905, 1913) and several North American and European bradoriid species. They also postulated the presence of an anterodorsal node, but we are unable to recognize this feature.

Our efforts at re-collecting *A. flexilis* from the type locality, together with an extensive search of the New Brunswick Museum collections from this locality, have not yielded additional specimens. *A. flexilis* must be considered a *nomen dubium*.

Occurrence. Type locality and stratum.

Genus ESCASONA Matthew, 1902

1902 *Escasona*, n gen., Matthew, p. 457.
1993*b* *Escasona* Matthew, 1902; Hinz-Schallreuter, p. 428.

Type species. By original designation; *Escasona rutellum* Matthew, 1902, p. 458, from the Mid Cambrian type Dugald Formation, Cape Breton, Nova Scotia.

Remarks. We regard *Escasona* as a *nomen dubium*. Of the four species which Matthew (1902) referred to it, three were each based on a single specimen from Dugald Brook, Nova Scotia: the type species, *E. rutellum* (holotype, ROM 155CM; Dugald Formation division E3f), has neither ornament, an admarginal ridge nor a well-defined dorsum and is probably a fragment of a non-bradoriid fossil; *Escasona*(?) *vetus* Matthew, 1902 (holotype, ROM 74CM; Dugald Formation division E1d) is an incomplete valve fragment, possibly referable to *Bradoria*; *Escasona*(??) *ingens* Matthew, 1902 (holotype, ROM 686CM; Eskasonie Formation) is a sheared, incomplete, indeterminate fossil. The fourth species, *Beyrichona ovata* Matthew, 1895 (holotype ROM 137CM; type Hanford Brook Formation), is an incomplete fragment of *Beyrichona tinea* (see above).

Order PHOSPHATOCOPIDA Müller, 1964

Remarks. For distinctions between the orders Phosphatocopida and Bradoriida see *Remarks* on the latter.

Family VESTROGOTHIIDAE Kozur, 1974

Genus CYCLOTRON Rushton, 1969

1902 *Polyphyma* gen. nov., Groom, pp. 83–88.
1969 *Cyclotron* nom. nov.; Rushton, p. 216.
1994*c* *Cyclotron* Rushton, 1969; Williams *et al.*, p. 126 (see for full discussion and synonymy).

Type species. By original designation; *Polyphyma lapworthi* Groom, 1902, from the Upper Cambrian of Hereford and Worcester, England.

Cyclotron lapworthi (Groom, 1902)

Plate 8, figures 8–9

1902 *Polyphyma lapworthi* sp. nov., Groom, pp. 83–88, pl. 3, figs 1–8.
1931*a* *Polyphyma marginata*, new species, Ulrich and Bassler, p. 66, pl. 8, figs 28–29.
1994*c* *Cyclotron lapworthi* (Groom, 1902); Williams *et al.*, p. 128, figs 1, 6 [see for full synonymy].

Lectotype. Designated Williams *et al.* 1994*c*, a left valve, OUM A.1336; thin shale unit, base of the late Cambrian White-Leaved-Oak Shales, northern end of Chase End Hill, southern Malvern Hills, England (UK Grid Ref.: SO 762 358).

Remarks. C. lapworthi was described fully by Williams *et al.* (1994*c*). It is a valuable marker fossil for the early Late Cambrian *Olenus* trilobite Biozone in Britain, Newfoundland and possibly Scandinavia.

Occurrence. Late Cambrian *Olenus* trilobite Biozone: Maclean Brook Formation at Manuels River, Conception Bay, eastern Newfoundland; thin shale unit at the base of the White-Leaved-Oak Shales, Malvern Hills, Hereford and Worcester and Outwood Shales Formation, Nuneaton, Warwickshire, southern Britain; possibly also from Öland and Jämtland, Sweden (see Williams *et al.* 1994*c*).

Genus VESTROGOTHIA Müller, 1964

Type species. By original designation; *Vestrogothia spinata* Müller, 1964, from the Upper Cambrian of Sweden.

Vestrogothia? *trituberculata* (Lochman and Hu, 1960)

Plate 8, figure 7

1960 *Dielymella*? *trituburculata* n. sp., Lochman and Hu, p. 826, pl. 98, fig. 56.
*non*1978 *Hesslandona trituberculata* (Lochman and Hu, 1960); Rushton, p. 279, pl. 26, fig. 11 [= indeterminate vestrogothiid juvenile, see Williams and Siveter in press].
1993*b* *Dielymella*? *trituberculata* Lochman and Hu, 1960; Hinz-Schallreuter, p. 396 [referred to *Hesslandona*].

Holotype. A small (0·91 mm long), juvenile left valve, USNM 138145 (Lochman and Hu, pl. 98, fig. 56; Pl. 8, fig. 7); Late Cambrian *Cedaria* trilobite Biozone, 27·5 metres below the top of the Du Noir Limestone, Sheep Mountain section, Northwest Wind River Mountains, north-eastern Sublette County, Wyoming.

Remarks. The type and only specimen has three dorsally aligned nodes characteristic of vestrogothiids (see Hinz-Schallreuter 1993*b*). Additional (adult) specimens are required in order to assess accurately the affinities of this species.

Occurrence. Type locality and stratum.

Genus Uncertain

'*Falites*' *pateli* Landing, 1980

1980 *Falites pateli* n. sp., Landing, p. 758, text-fig. 4A–E.
1993*b* *Falites pateli* Landing, 1980; Hinz-Schallreuter, p. 400.

Holotype. A small (juvenile?) carapace, USNM 256963 (Landing 1980, text-fig. 4A, C); Upper Cambrian, Chesley Drive Group, New Brunswick.

Remarks. Landing's (1980) figured specimens are incomplete, probably small juveniles. The elongate carapace and small size and shape of the anterodorsal(?) node recall vestrogothiids (see Williams *et al.* 1994*b*, 1994*c*).

Occurrence. Late Cambrian *Parabolina spinulosa* trilobite Biozone, lower part of the Chesley Drive Group, New Brunswick.

EXPLANATION OF PLATE 9

Figs 1–8. *Dielymella recticardinalis* Ulrich and Bassler, 1931*a*; Mid Cambrian Bright Angel Shale, Bass Canyon, Grand Canyon, Arizona. 1, USNM 487777; mostly exfoliated, posteriorly incomplete right valve (Ulrich and Bassler 1931*a*, pl. 10, fig. 4), stereo-pair; ×6. 2, USNM 487776; partly exfoliated right valve, lectotype (Ulrich and Bassler 1931*a*, pl. 10, fig. 3), stereo-pair; ×6. 3–4, USNM 487779; partly exfoliated carapace (Ulrich and Bassler 1931*a*, pl. 10, fig. 7); ×6: 3, dorsal view, stereo-pair. 4, left valve. 5, USNM 56511; left valve of partly exfoliated carapace (holotype of *Dielymella recticardinalis angustata* Ulrich and Bassler, 1931*a*, pl. 10, fig. 8); ×8. 6, USNM 487775; mostly exfoliated, posteriorly incomplete left valve (lectotype of *Dielymella nasuta* Ulrich and Bassler, 1931*a*, pl. 10, fig. 10); ×6·5. 7, USNM 56508; exfoliated, partly open carapace (paralectotype of *Dielymella nasuta* Ulrich and Bassler, 1931*a*, pl. 10, fig. 11); ×5·5. 8, USNM 487778; right valve of partly exfoliated carapace (Ulrich and Bassler 1931*a*, pl. 10, fig. 5); ×6.
Fig. 9. *Dielymella*? *brevis* Ulrich and Bassler, 1931*a*; ?early Cambrian sandstone, Sunset Hill, near Lake Dunmore, Vermont; USNM 487781; abraded internal mould of posteriorly incomplete ?right valve, lectotype (Walcott 1890, pl. 80, fig. 4*a*); ×3·5.
All are lateral views except fig. 3.

SIVETER and WILLIAMS, *Dielymella*, *Dielymella*?

Order and Family Uncertain

Genus DIELYMELLA Ulrich and Bassler, 1931*a*

Type species. By original designation; *Dielymella recticardinalis* Ulrich and Bassler, 1931*a*, from the Middle Cambrian of the Grand Canyon, Arizona.

Diagnosis. Large (adults *c.* 10 mm long), elongate, preplete, equivalved carapace. Dorsal margin straight, hinge-line well developed. Short, stout, anterocardinal spine. Valves punctate.

Remarks. The systematic position of *Dielymella* is uncertain. The short, presumed anterocardinal spine often appears to have been broken and may have been longer. Several similar-sized Cambrian bivalved arthropods have comparable spines (see Siveter *et al.* 1996; Williams *et al.* 1996) but their systematic affinites are likewise uncertain. If our orientation of the carapace is correct, *Dielymella*, although large, shows the preplete shape typical of many Phosphatocopida.

Indiana impressa Ulrich and Bassler, 1931*a*, from the Mid Cambrian Bright Angel Shale of Arizona, is based on several exfoliated and incomplete juveniles (USNM 56463), which are possibly referable to *Dielymella*.

Occurrence. Middle Cambrian of Arizona and ?Lower Cambrian of Vermont (Ulrich and Bassler 1931*a*).

Dielymella recticardinalis Ulrich and Bassler, 1931*a*

Plate 9, figures 1–8

1931*a*	*Dielymella recticardinalis*, new species, Ulrich and Bassler, p. 86, pl. 10, figs 3–7.
1931*a*	*Dielymella recticardinalis angustata*, new variety, Ulrich and Bassler, p. 87, pl. 10, fig. 8.
1931*a*	*Dielymella nasuta*, new species, Ulrich and Bassler, p. 88, pl. 10, figs 10–11.
1931*a*	*Dielymella appressa*, new species, Ulrich and Bassler, p. 88, pl. 10, fig. 9.
1931*a*	*Dielymella dorsalis*, new species, Ulrich and Bassler, p. 89, pl. 10, fig. 1.
1945	*Dielymella appressa* Ulrich and Bassler; Resser, p. 216, pl. 27, figs 10, 21.
1945	*Dielymella dorsalis* Ulrich and Bassler; Resser, p. 216, pl. 26, fig. 20.
1945	*Dielymella nasuta* Ulrich and Bassler; Resser, p. 216, pl. 27, figs 11–12, 23.
1945	*Dielymella recticardinalis* Ulrich and Bassler; Resser, p. 216, pl. 27, figs 13, 15, 19–20.
1945	*Dielymella recticardinalis angustata* Ulrich and Bassler; Resser, p. 217, pl. 27, fig. 24.
1961	*D. recticardinalis*; Sylvester-Bradley, p. Q103, fig. 39.5.

Lectotype. Designated herein; a partly exfoliated adult right valve, USNM 487776 (Ulrich and Bassler 1931*a*, pl. 10, fig. 3; Pl. 9, fig. 2); Mid Cambrian *Glossopleura* trilobite Biozone, Bright Angel Shale, Grand Canyon, Arizona.

Material. Some 23 specimens: USNM 56505–56506, 56508, 56510–56511, 487775–487780.

Measurements. Specimens from the Bright Angel Shale are 4·7–11·0 mm long.

Diagnosis. Dielymella species with anterocardinal corner having short, stout, spine.

Description. Large (adults *c.* 10–11 mm), elongate, preplete, equivalved, bivalved carapace; no posterior gape apparent. Dorsal margin straight, shorter than valve length. Hinge-line well developed. Short, stout anterocardinal spine. Valve surface gently convex and finely punctate. Anterior cardinal angle much greater than posterior cardinal angle. Ventral part of lateral outline evenly convex. Anterodorsal part of lateral outline almost straight. No lobation, nodation or marginal structures.

Remarks. We consider *Dielymella recticardinalis angustata*, *D. appressa*, *D. nasuta* and *D. dorsalis* (all Ulrich and Bassler, 1931*a*) to be synonymous with *D. recticardinalis*. Syntypes of *D. recticardinalis* comprise eight specimens (*ex* USNM 56510): the lectotype (USNM 487776); an incomplete adult right valve (USNM 487777; Ulrich and Bassler 1931*a*, pl. 10, fig. 4; Pl. 9, fig. 1); two partly exfoliated juvenile carapaces (USNM 487778–487779; Ulrich and Bassler 1931*a*, pl. 10, figs 5, 7; Resser 1945, pl. 27, figs 13, 15; Pl. 9, figs 3–4, 8); and four other valves and carapaces (which retain the number USNM 56510).

Specimens referred to *D. appressa* include 11 valves and carapaces (USNM 56506; Ulrich and Bassler 1931*a*, pl. 10, fig. 9; Resser 1945, pl. 27, figs 10, 21), one of which has a thick shell (renumbered USNM 487780). *D. recticardinalis angustata* is based on a juvenile carapace (holotype, USNM 56511; Ulrich and Bassler 1931*a*, pl. 10, fig. 8; Resser 1945, pl. 27, fig. 24; Pl. 9, fig. 5).

Syntypes of *D. nasuta* are an apparently hingeless or badly deformed specimen of uncertain affinity (USNM 56509), a juvenile carapace (USNM 56508; Ulrich and Bassler 1931*a*, pl. 10, fig. 11; Resser 1945, pl. 27, figs 11–12; Pl. 9, fig. 7) and a posteriorly incomplete adult left valve (herein designated lectotype, USNM 487775, *ex* USNM 56508; Ulrich and Bassler 1931*a*, pl. 10, fig. 10; Resser 1945, pl. 27, fig. 23; Pl. 9, fig. 6). *D. dorsalis* is based on a juvenile (holotype, USNM 56505; Ulrich and Bassler 1931, pl. 10, fig. 1; Resser 1945, pl. 26, fig. 20).

Occurrence. Mid Cambrian *Glossopleura* trilobite Biozone, Bright Angel Shale, Nunkoweap Valley and Bass Canyon, Grand Canyon, northern Arizona.

Dielymella? *brevis* Ulrich and Bassler, 1931a

Plate 9, figure 9

?1884 *Nothozoe vermontana*, n. sp., Whitfield, p. 144, pl. 14, figs 14–15.
1890 *Nothozoe*? *vermontana* Whitfield; Walcott, p. 628, pl. 80, fig. 4*a–b*, ?fig. 4.
1931*a* *Dielymella brevis*, new species, Ulrich and Bassler, p. 89, pl. 10, figs 12–13.

Lectotype. Designated herein, USNM 487781 (Walcott 1890, pl. 80, fig. 4*a*; Ulrich and Bassler 1931*a*, pl. 10, fig. 12; Pl. 9, fig. 9), abraded internal mould of a posteriorly incomplete right? valve; from a loose boulder of ?early Cambrian Sandstone, Sunset Hill, Vermont.

Material and measurements. Two specimens: USNM 56504 (15 mm long) and USNM 487781 (16 mm long).

Remarks. Of three specimens referred to *Nothozoe*? *vermontana* Whitfield by Walcott (1890), Ulrich and Bassler (1931*a*) referred two (Walcott 1890, pl. 80, fig. 4*a–b*) to their species *Dielymella brevis*. We consider that both are too poorly preserved to be referred to *Dielymella* with certainty. The affinity of Walcott's third specimen (1890, pl. 80, fig. 4), which appears to have a markedly dissimilar lateral outline, is also uncertain.

The affinities of *Nothozoe vermontana* Whitfield, 1884, from the Lower Cambrian near Lake Dunmore, Vermont, need further investigation.

Occurrence. Loose boulder of ?early Cambrian sandstone, Sunset Hill, Vermont. Walcott (1890, p. 628) apparently discovered additional specimens associated with *Olenellus* and *Hytolithes*, from exposures east of Bennington, Vermont, but there is no record of such material in the USNM.

CONCLUSIONS

We conclude that the Cambrian Bradoriida and Phosphatocopida of North America: (1) are recovered mostly as 'crack-out' material from clastic rocks (acid-resistant valves, from limestones, are a minor component of the known fauna); (2) are most abundant in the Canadian Atlantic Provinces, British Columbia, Arizona, New York State and Tennessee; (3) contain representatives of the families Bradoriidae (junior synonym: Indianidae), Beyrichonidae, Hipponicharionidae, Svealutidae and Vestrogothiidae, and that *Escasona* and the previously widely recorded *Aluta* are *nomina dubia*; (4) comprise 16 genera and 26 species (including taxa described under open nomenclature) and that faunal diversity is therefore considerably less than indicated in previous records (about 100 species); (5) include forms with a phosphatic carapace (e.g. *Beyrichona*,

Hipponicharion, *Indiana*, *Liangshanella*, *Petrianna*) and, therefore, carapace composition appears to be an unreliable criterion for distinguishing bradoriids from phosphatocopids; (6) include some taxa provincial to palaeocontinental Avalonia (*Beyrichona*, *Cyclotron*, *Hipponicharion*) and Laurentia (*Dielymella*, *Walcottella*), thus supporting the notion of the presence of an intervening (Iapetus) Ocean; (7) occur in the Lower, Middle and Upper Cambrian, in both Avalonia and Laurentia; (8) consist mainly of short-ranging species, most of which currently have only local occurrence and a few of which have international correlative value; (9) occur as low diversity marine assemblages; (10) lived in shallow cratonic and nearshore shelf (e.g. Arizona; Nova Scotia, New Brunswick) and deeper offshore shelf (Newfoundland, British Columbia, Tenneessee) areas; the deeper settings on Laurentia have a few cosmopolitan bradoriids of Asiatic (*Liangshanella* and cambriids) and Australian (*Indota*) affinities; (11) are facies related and were probably mostly benthic or nektobenthic.

Acknowledgements. We gratefully acknowledge the support of the Natural Environment Research Council (Grant GR3/8655). Prof. R. A. Robison (University of Kansas), Mrs J. Waddington (Royal Ontario Museum), Drs R. F. Miller (New Brunswick Museum), Derek J. Siveter (Oxford University Museum), R. H. Benson, J. Thompson and Mr M. Florence (Smithsonian Institution) kindly arranged loan of museum collections. Mr M. Florence also provided invaluable, previously unpublished information about the Washington collections. We thank Prof. R. F. Lundin (Arizona State University) and Dr A. W. A. Rushton (British Geological Survey) for much valued reviews of our manuscript. We are grateful to Drs R. F. Miller and I. Patel (University of New Brunswick) for hospitality and valuable field assistance in New Brunswick. DJS gratefully acknowledges a G. F. Matthew Research Grant. Prof. Hou Xianguang (Academica Sinica, Nanjing) and Drs L. Melnikova (Palaeontological Institute, Moscow) and V. Berg-Madsen (Uppsala University) facilitated access to comparative Chinese, Russian and Scandinavian material. We thank the *Eskasonie* Indian Band Office (Eskasonie, Nova Scotia) for permission to visit localities on their land.

REFERENCES

ABUSHIK, A. F. 1960. [First discovery of leperditaceans from the Cambrian of the Siberian Platform.] *Vestnik Leningradskogo Universitet, Seriya Geologii i Geografii*, **6**, 93–98. [In Russian].

ANDRES, D. 1969. Ostracoden aus dem mittleren Kambrium von Öland. *Lethaia*, **2**, 165–180.

BERDAN, J. M. 1984. Leperditicopid ostracodes from Ordovician rocks of Kentucky and nearby states and characteristic features of the Order Leperditicopida. *Professional Paper of the United States Geological Survey*, **1066-J**, 1–40, pls 1–11.

BEUS, S. S. 1987. Geology along the South Kaibab Trail, eastern Grand Canyon, Arizona. 371–378. *In* BEUS, S. S. (ed.). *Rocky Mountain Section of the Geological Society of America Centennial Field Guide*, **2**, xx+475 pp.

BOLTON, T. E. and COPELAND, M. J. 1963. *Cambrotrypa* and *Bradoria* from the middle Cambrian of western Canada. *Journal of Paleontology*, **37**, 1069–1070, pl. 143.

BRASIER, M. D. and COWIE, J. W. 1989. Other areas: North-west Canada; California, Nevada, and Mexico; Morocco, Spain, and France. 105–114. *In* COWIE, J. W. and BRASIER, M. D. (eds). *The Precambrian–Cambrian boundary*. Oxford Monographs on Geology and Geophysics, 12. Clarendon Press, Oxford, ix+213 pp.

—— —— and TAYLOR, M. E. 1994. Decision on the Precambrian-Cambrian Boundary Stratotype. *Episodes*, **17**, 3–8.

BULMAN, O. M. B. and RUSHTON, A. W. A. 1973. Tremadoc faunas from boreholes in Central England. *Bulletin of the Geological Survey of Great Britain*, **43**, 1–41, pls 1–7.

CHAPMAN, F. 1918. Ostracoda from the upper Cambrian limestone of South Australia. *Proceedings of the Royal Society of Victoria, New Series*, **31**, 108–174, pl. 9.

COBBOLD, E. S. 1921. The Cambrian horizons of Comley (Shropshire) and their Brachiopoda, Pteropoda, Gasteropoda, etc. *Quarterly Journal of the Geological Society, London*, **76** (for 1920), 325–386, pls 21–24.

—— 1934. *In* COBBOLD, E. S. and POCOCK, R. W. The Cambrian area of Rushton (Shropshire). *Philosophical Transactions of the Royal Society, Series B*, **223**, 305–409, pls 38–45.

—— 1936. The Conchostraca of the Cambrian area of Comley, Shropshire, with a note on a new variety of *Atops reticulatus* (Walcott). *Quarterly Journal of the Geological Society, London*, **92**, 221–235, pls 14–15.

CONWAY MORRIS, S. 1986. The community structure of the middle Cambrian Phyllopod Bed (Burgess Shale). *Palaeontology*, **29**, 423–467.

—— 1989. South-eastern Newfoundland and adjacent areas (Avalon Zone). 7–39. *In* COWIE, J. W. and BRASIER, M. D. (eds). *The Precambrian–Cambrian boundary*. Oxford Monographs on Geology and Geophysics, 12. Clarendon Press, Oxford, ix+213 pp.

COPELAND, M. J. 1986. *Bullaluta kindlei* n. gen., n. sp. (Ostracoda, Archaeocopida) from Zone 5 (Late Cambrian, *Cedaria-Crepicephalus*) of the Cow Head Group, western Newfoundland. 339–403. *In* Current Research, Part B. *Geological Survey of Canada, Paper* 86–1*B*.

FORD, S. W. 1873. Descriptions of new species of fossils from the Lower Potsdam Group at Troy, New York. *American Journal of Science, Series* 3, **6**, 137–140.

FREDERICKSON, E. A. 1946. A Cambrian ostracode from Oklahoma. *Journal of Paleontology*, **20**, 578.

GROOM, T. 1902. On *Polyphyma*, a new genus belonging to the Leperditidae, from the Cambrian shales of Malvern. *Quarterly Journal of the Geological Society, London*, **58**, 83–88, pl. 3.

GRÜNDEL, J. 1981. Die Bradoriida. 57–73, pls 1–3. *In* GRÜNDEL, J. and BUCHHOLZ, A. Bradoriida aus kambrischen Geschieben vom Gebiet der nördlichen DDR. *Freiburger Forschungshefte, Series C*, **363**.

GÜRICH, G. 1929. *Silesicaris* von Leipe und die Phyllokariden überhaupt. *Mineralogisch-geologischen Staatsinstitut, Hamburg, Mitteilungen*, **11**, 21–90.

HAYES, A. O. and HOWELL, B. F. 1937. Geology of Saint John, New Brunswick. *Special Paper of the Geological Society of America*, **5**, ix+146 pp.

HINTZE, L. F. and ROBISON, R. A. 1975. Middle Cambrian stratigraphy of the House, Wah Wah, and adjacent ranges in western Utah. *Bulletin of the Geological Society of America*, **86**, 881–891.

HINZ, I. C. U. 1992. On *Pejonesia sestina* (Fleming). *A Stereo-Atlas of Ostracod Shells*, **19**, 5–8.

—— 1993. Evolutionary trends in archaeocopid ostracodes. 3–12. *In* McKENZIE, K. G. and JONES, P. J. *Ostracoda in the life sciences*. Proceedings of the 11th International Symposium on Ostracoda, Warrnambool, Australia. Balkema, Rotterdam.

HINZ-SCHALLREUTER, I. 1993*a*. Ostracodes from the middle Cambrian of Australia. *Neues Jahrbuch für Geologie und Paläontologie, Abhandlungen*, **188**, 305–326.

—— 1993*b*. Cambrian ostracodes mainly from Baltoscandia and Morocco. *Archiv für Geschiebekunde*, **1**, 385–448.

—— and JONES, P. J. 1994. *Gladioscutum lauriei* n. gen. n. sp. (Archaeocopida) from the middle Cambrian of the Georgina Basin, central Australia. *Paläontologische Zeitschrift*, **68**, 361–375.

HOU XIANGUANG 1987. Early Cambrian large bivalved arthropods from Chengjiang, eastern Yunnan. *Acta Palaeontologica Sinica*, **26**, 286–298, pls 1–6. [In Chinese and English].

—— and BERGSTRÖM, J. 1991. The arthropods of the Lower Cambrian Chengjiang fauna, with relationships and evolutionary significance. 179–187. *In* SIMONETTA, A. and CONWAY MORRIS, S. (eds). *The early evolution and the significance of problematic taxa*. Cambridge University Press, Cambridge, ix+296 pp.

—— SIVETER, DAVID J., WILLIAMS, M., WALOSSEK D. and BERGSTRÖM, J. 1996. Appendages of the arthropod *Kunmingella* from the early Cambrian of China: its bearing on the systematic position of the Bradoriida and the fossil record of the Ostracoda. *Philosophical Transactions of the Royal Society, Series B*, **351**, 1131–1145.

HUO SHICHENG 1956. [Brief notes on lower Cambrian Archaeostraca from Shensi and Yunnan.] *Acta Palaeontologica Sinica*, **4**, 425–445, pls 1–3. [In Chinese].

—— and SHU DEGAN 1985. *Cambrian Bradoriida of South China*. Northwest University Publishing House, Xi'an, 252 pp., 37 pls. [In Chinese, with English summary].

—— —— and CUI ZHILIN 1991. [*Cambrian Bradoriida of China*.] Geological Publishing House, Beijing, 249 pp., 46 pls. [In Chinese, with English summary].

—— —— ZHANG XIGUANG, CUI ZHILIN and TONG HAOWEN 1983. Notes on Cambrian bradoriids from Shaanxi, Yunnan, Sichuan, Guizhou, Hubei and Guangdong. *Journal of Northwest University*, **13**, 56–75, pls 1–6. [In Chinese, with English abstract].

HUTCHINSON, R. D. 1952. The stratigraphy and trilobite faunas of the Cambrian sedimentary rocks of Cape Breton Island, Nova Scotia. *Memoir of the Geological Society of Canada*, **263**, v+124 pp., pls 1–7.

IVANOVA, V. A. 1960. [On the origin and phylogeny of the Ostracodoidea.] *Paleontologicheskiy Zhurnal*, **1960** (3), 21–27. [In Russian].

JONES, P. J. and McKENZIE, K. G. 1980. Queensland middle Cambrian Bradoriida (Crustacea): new taxa, palaeobiogeography and biological affinities. *Alcheringa*, **4**, 203–225.

JONES, T. R. 1881. Notes on the Palaeozoic bivalved Entomostraca. No. 12. Some Cambrian and Silurian Leperditiae and Primitiae. *Annals and Magazine of Natural History, Series 5*, **8**, 332–350, pls 19–20.

KEMPF, E. K. 1986. *Index and bibliography of marine Ostracoda 1, Index A*. Sonderveröffentlichungen Geologisches Institut der Universität zu Köln, 50, 762 pp.

KINDLE, C. H. 1981. Cambrian faunas in limestone conglomerates of western Newfoundland. 106–110. In *United States Department of the Interior, Geological Survey, Open File Report*, 81–743.

KOBAYASHI, T. and KATO, F. 1951. On the ontogeny and the ventral morphology of *Redlichia chinensis* with description of *Alutella nakamurai* new gen. and sp. *Journal of the Faculty of Science, University of Tokyo, Section 2*, **8**, 85–90.

KOZUR, H. 1974. Die Bedeutung der Bradoriida als Vorläufer der postkambrischen Ostracoden. *Zeitschrift für Geologische Wissenschaften*, **2**, 823–830.

KUMMEROW, E. 1931. Über die unterschiede zwischen Phyllocariden und ostracoden. *Centralblatt für Mineralogie, Geologie und Paläontologie, Abteilung B*, **1931**, 242–257.

LANDING, E. 1980. Late Cambrian–early Ordovician macrofaunas and phosphatic microfaunas, St. John Group, New Brunswick. *Journal of Paleontology*, **54**, 752–761.

—— 1991. Upper Precambrian through lower Cambrian of Cape Breton Island: faunas, paleoenvironments and stratigraphic revision. *Journal of Paleontology*, **65**, 570–595.

—— 1992. Lower Cambrian of southeastern Newfoundland. Epeirogeny and Lazarus faunas, lithofacies-biofacies linkages, and the myth of a global chronostratigraphy. 283–309. *In* LIPPS, J. H. and SIGNOR, P. W. (eds). *Origin and early evolution of the Metazoa*. Plenum Press, New York, 570 pp.

—— 1995. Upper Placentian–Branchian series of mainland Nova Scotia (middle-upper lower Cambrian): faunas, paleoenvironments and stratigraphic revision. *Journal of Paleontology*, **69**, 475–495.

—— 1996. Avalon: insular continent by the latest Precambrian. *Special Paper of the Geological Society of America*, **304**, 29–63.

—— and WESTROP, S. R. 1996. Upper Lower Cambrian depositional sequence in Avalonian New Brunswick. *Canadian Journal of Earth Sciences*, **33**, 404–417.

LAURENCE, R. A. and PALMER, A. R. 1963. Age of the Murray Shale and Hesse Quartzite on Chilhowee Mountain, Blount County, Tennessee. *Professional Paper of the United States Geological Survey*, **475-C**, 53–54.

LINNARSSON, J. G. O. 1869. Om Vestergötlands Cambriska och Siluriska Aflagrigar. *Kongliga Svenska Vetenskaps-Akademiens Handlingar*, **8**, 1–89, pls 1–2.

LOCHMAN, C. 1956. Stratigraphy, paleontology, and paleogeography of the *Elliptocephala asaphoides* strata in Cambridge and Hoosick quadrangles, New York. *Bulletin of the Geological Society of America*, **67**, 1331–1396, pls 1–8.

—— and HU CHUNG-HUNG 1960. Upper Cambrian faunas from the Northwest Wind River Mountains, Wyoming. Part 1. *Journal of Paleontology*, **34**, 793–834, pls 95–100.

LOCHMAN-BALK, C. 1971. The Cambrian of the craton of the United States. 79–168. *In* HOLLAND, C. H. (ed.). *Lower Palaeozoic rocks of the world. Volume 1: Cambrian of the New World*. Wiley-Interscience, London, 456 pp.

MALZ, H. 1990. Nomenclatorial annotations on 'Cambrian and early Ordovician "Ostracoda" (Bradoriida) in China' by Shu. *Courier Forschungs-Institut Senckenberg*, **123**, 331–332.

MATTHEW, G. F. 1886. Illustrations of the fauna of the St. John Group continued. No. 3: Descriptions of new genera and species. *Proceedings and Transactions of the Royal Society of Canada, Series 1*, **3** (for 1885), 29–84, pls 5–7.

—— 1890. On Cambrian organisms in Acadia. *Proceedings and Transactions of the Royal Society of Canada, Series 1*, **7** (for 1889), 135–161, pls 5–9.

—— 1891. Notes on Cambrian faunas. The Taconic fauna of Emmons compared with Cambrian horizons of the St. John Group. *American Geologist*, **8**, 287–291.

—— 1894. Illustrations of the fauna of the St. John Group, No. 8. *Proceedings and Transactions of the Royal Society of Canada, Series 1*, **11** (for 1893), 85–129, pls 16–17.

—— 1895. The *Protolenus* Fauna. *Transactions of the New York Academy of Sciences*, **14** (for 1894–95), 101–153, pls 1–11.

—— 1896. Faunas of the *Paradoxides* Beds in eastern North America, No. 1. *Transactions of the New York Academy of Sciences*, **15**, 192–247, pls 14–17.

—— 1898. Studies on Cambrian faunas, No. 2. The Cambrian System in the Kennebacasis Valley. *Transactions of the Royal Society of Canada, Series* 2, **4** (for 1897), 123–153, pls 1–2.

—— 1899. Preliminary notice of the Etcheminian Fauna of Cape Breton. *Bulletin of the Natural History Society of New Brunswick*, **4**, 198–208, pls 1–4.

—— 1902. Ostracoda of the basal Cambrian rocks in Cape Breton. *Canadian Record of Science*, **8**, 437–470, pls 1–2.

—— 1903. *Report on the Cambrian Rocks of Cape Breton*. Geological Survey of Canada, Ottawa, 246 pp., 18 pls.

—— 1906. Notes on Cambrian faunas, no. 9 – Ostracoda. *Bulletin of the Natural History Society of New Brunswick*, **5**, 406, 475–480.

MARTIN, F. and DEAN, W. T. 1988. Middle and upper Cambrian acritarch and trilobite zonation at Manuels River and Random Island, Eastern Newfoundland. *Bulletin of the Geological Survey of Canada*, **381**, 1–91, pls 1–18.

McCUTCHEON, S. R. 1987. Cambrian stratigraphy in the Hanford Brook area, southern New Brunswick, Canada. 399–402. *In* ROY, D. C. (ed.). *Northeastern Section of the Geological Society of America Centennial Field Guide*, **5**, xxiii + 481 pp.

McKERROW, W. S., SCOTESE, C. R. and BRASIER, M. D. 1992. Early Cambrian continental reconstructions. *Journal of the Geological Society, London*, **149**, 599–606.

McMENAMIN, M. A. S. and McMENAMIN, D. L. S. 1990. *The emergence of animals, the Cambrian breakthrough*. Columbia University Press, New York, x + 217 pp.

MELNIKOVA, L. M. 1983. [Ostracoda (Bradoriida).] *In* GRIGOREVA, H. B., MELNIKOVA, L. M. and PELMAN, YU. L. [Brachiopoda, Ostracoda (Bradoriida) and problematica from the stratotype region of the Lower Cambrian stages.] *Paleontologicheskiy Zhurnal*, **1983**, 54–58, pl. 5. [In Russian].

—— 1984. [Bradoriida.] *In* KHAZANOVICH, K. K., POPOV, L. E. and MELNIKOVA, L. M. [Brachiopoda, Ostracoda (Bradoriida) and Hyolithelminta from the Sablinsk Beds of the Leningrad region.] *Paleontologicheskiy Zhurnal*, **1984**, 33–47, pl. 3. [In Russian].

—— 1987. [Bradoriids from the Tikres Formation (Lower Cambrian) in Estonia.] *Paleontologicheskiy Zhurnal*, **1987**, 128–131. [In Russian].

—— 1988. [Some bradoriids (Crustacea) from the Botomian Stage of the East Trans-Baikal Region.] *Paleontologicheskiy Zhurnal*, **1988** (1), 114–117. [In Russian].

—— 1992. [The first late Cambrian bradoriids (Ostracoda) from Gorny Altay.] *Paleontologicheskiy Zhurnal*, **1992**(4), 80–82. [In Russian].

—— SIVETER, DAVID J. and WILLIAMS, M. in press. Cambrian Bradoriida and Phosphatocopida (Arthropoda) of the former Soviet Union. *Journal of Micropalaeontology*.

MILLER, R. F. 1988. Catalogue of type fossils in the New Brunswick Museum. *The New Brunswick Museum, Publications in Natural Science*, **7**, 1–75.

—— and BUHAY, D. N. 1990. Life and letters of George Frederic Matthew: geologist and palaeontologist. *The New Brunswick Museum, Publications in Natural Science*, **8**, 1–89.

MILLS, A. A. 1988. Silver as a removable conductive coating for scanning electron microscopy. *Scanning Microscopy*, **2**, 1265–1271.

MOBERG, J. C. and SEGERBERG, C. O. 1906. Bidrag till Kännedomen om Ceatopygeregionen med särskild hänsyn till dess utveckling i Fogelsångstrakten. *Acta Universitatis Lundensis, Nova Series, Afdeling 2*, **2** (7), 1–116, pls 1–7.

MÜLLER, K. J. 1964. Ostracoda (Bradorina) mit phosphatischen Gehäusen aus dem oberkambrium von Schweden. *Neues Jahrbuch für Geologie und Paläontologie, Abhandlungen*, **121**, 1–46, pls 1–5.

—— 1979. Phosphatocopine ostracodes with preserved appendages from the Cambrian of Sweden. *Lethaia*, **12**, 1–27.

—— 1982. *Hesslandona unisulcata* sp. nov. with phosphatized appendages from upper Cambrian 'orsten' of Sweden. 276–307. *In* BATE, R. H., ROBINSON, E. and SHEPPARD, L. M. (eds). *Fossil and Recent Ostracoda*. Ellis Horwood, Chichester, 492 pp.

—— and WALOSSEK, D. 1991. Ein Blick durch das Orsten-Fenster in die Arthropodenwelt vor 500 Millionen Jahren. *Verhandlungen der Deutschen Zoologischen Gesellschaft*, **84**, 281–294.

NORTH, F. K. 1971. The Cambrian of Canada and Alaska. 219–324. *In* HOLLAND, C. H. (ed.). *Lower Palaeozoic rocks of the world. Volume 1: Cambrian of the New World*. Wiley-Interscience, London, 456 pp.

ÖPIK, A. A. 1961. The geology and palaeontology of the headwaters of the Burke River, Queensland. *Bulletin of the Bureau of Mineral Resources, Geology and Geophysics*, **53**, 1–249, pls 1–24.

—— 1968. Ordian (Cambrian) Crustacea Bradoriida of Australia. *Bulletin of the Bureau of Mineral Resources, Geology and Geophysics*, **103**, 1–46, pls 1–4.

PALMER, A. R. 1954. The faunas of the Riley Formation in central Texas. *Journal of Paleontology*, **28**, 709–786, pls 76–92.

—— 1971. The Cambrian of the Great Basin and adjacent areas, western United States. 1–78. *In* HOLLAND, C. H. (ed.). *Lower Palaeozoic rocks of the world. Volume 1: Cambrian of the New World*. Wiley-Interscience, London, 456 pp.

POULSEN, C. 1932. The Lower Cambrian faunas of East Greenland. *Meddelelser om Grønland*, **87**, 6, 1–66.

PICKERING, K. T. and SMITH, A. G. 1995. Arcs and back-arc basins in the Early Palaeozoic Iapetus Ocean. *The Island Arc*, **4**, 1–67.

RAYMOND, P. E. 1935. *Leanchoilia* and other mid-Cambrian Arthropoda. *Bulletin of the Museum of Comparative Zoology, Harvard University*, **76**, 205–230.

RESSER, C. E. 1938. Cambrian System (restricted) of the southern Appalachians. *Special Paper of the Geological Society of America*, **15**, 1–140, pls 1–16.

—— 1945. Cambrian history of the Grand Canyon region, Part 2. Cambrian fossils of the Grand Canyon. *Publication of the Carnegie Institution of Washington*, **563**, 171–220, pls 16–27.

—— and HOWELL, B. J. 1938. Lower Cambrian *Olenellus* Zone of the Appalachians. *Bulletin of the Geological Society of America*, **49**, 195–248.

ROBISON, R. A. 1984. Cambrian Agnostida of North America and Greenland; Part I, Ptychagnostidae. *Paleontological Contributions of the University of Kansas*, *Paper* **109**, 1–59.

—— 1989. Middle Cambrian biotic diversity: examples from four Utah Lagerstätten. 77–93. *In* SIMONETTA, A. and CONWAY MORRIS, S. (eds). *The early evolution and the significance of problematic taxa*. Cambridge University Press, Cambridge, ix + 296 pp.

RUSHTON, A. W. A. 1966. The Cambrian trilobites from the Purley Shales of Warwickshire. *Monograph of the Palaeontographical Society*, **120** (511), 1–55, pls 1–6.

—— 1969. *Cyclotron*, a new name for *Polyphyma* Groom. *Geological Magazine*, **106**, 216–217.

—— 1978. Fossils from the middle-upper Cambrian transition in the Nuneaton district. *Palaeontology*, **21**, 245–283, pls 24–26.

—— SIVETER, DAVID J. and WILLIAMS, M. in press. The Cambrian. *In* KEEN, M. and WILKINSON, I. (eds). *A stratigraphical index of British Ostracoda*. British Micropalaeontological Society Series. Chapman and Hall, London.

SCOTESE, C. R. and McKERROW, W. S. 1990. Revised world maps and introduction. 1–21. *In* McKERROW, W. S. and SCOTESE, C. R. (eds). Palaeozoic palaeogeography and biogeography. *Memoir of the Geological Society, London*, **12**, 1–435.

SHU DEGAN 1990*a*. *Cambrian and Ordovician Bradoriida from Zhejiang, Hunan and Shaanxi provinces*. Northwest University Press, Xian. 95 pp., 20 pls. [In Chinese with English summary].

—— 1990*b*. Cambrian and early Ordovician 'Ostracoda' (Bradoriida) in China. *Courier Forschungs-Institut Senckenberg*, **123**, 315–330, pls 1–3.

SIVETER, DAVID J. 1984. Habitats and modes of life of Silurian ostracodes. 71–85. *In* BASSETT, M. G. (ed.). Autecology of Silurian organisms. *Special Papers in Palaeontology*, **32**, 1–295.

—— and WILLIAMS, M. 1995. An early Cambrian assignment for the Caerfai Group of South Wales. *Journal of the Geological Society, London*, **152**, 221–224.

—— —— ABUSHIK, A. F., BERG-MADSEN, V. and MELNIKOVA, L. 1993. On *Anabarochilina primordialis* (Linnarsson). *A Stereo-Atlas of Ostracod Shells*, **20**, 71–76.

—— —— and MELNIKOVA, L. 1994*a*. On *Cambria melnikovae* Ivanova. *A Stereo-Atlas of Ostracod Shells*, **21**, 13–16.

—— —— and BERG-MADSEN, V. 1994*b*. On *Wimanicharion matthewi* (Wiman). *A Stereo-Atlas of Ostracod Shells*, **21**, 17–20.

—— —— PEEL, J. S. and SIVETER, DEREK, J. 1996. Bradoriida (Arthropoda) from the early Cambrian of North Greenland. *Transactions of the Royal Society of Edinburgh: Earth Sciences*, **86** (for 1995), 113–121.

SIVETER, DEREK J. 1990. Photography. 505–508. *In* BRIGGS, D. E. G. and CROWTHER, P. R. (eds). *Palaeobiology: a synthesis*. Blackwell, Oxford, xiii + 583 pp.

STUBBLEFIELD, C. J. 1933. Notes on the fossils. *In* SMITH, S. On the occurrence of Tremadoc Shales in the Tortworth Inlier (Gloucestershire). *Quarterly Journal of the Geological Society, London*, **89**, 357–375.

SYLVESTER-BRADLEY, P. C. 1961. Archaeocopida. 100–105. *In* MOORE, R. C. (ed.). *Treatise on invertebrate paleontology. Part Q. Arthropoda 3*. Geological Society of America and University of Kansas Press, Boulder, Colorado and Lawrence, Kansas, xxiii + 442 pp.

TANOLI, S. K. and PICKERILL, R. K. 1988. Lithostratigraphy of the Cambrian–Lower Ordovician Saint John Group, southern New Brunswick. *Canadian Journal of Earth Sciences*, **25**, 669–690.

TAYLOR, K. and RUSHTON, A. W. A. 1971. The Pre-Westphalian geology of the Warwickshire Coalfield. *Bulletin of the Geological Survey of Great Britain*, **35**, 1–152, pls 1–12.

ULRICH, E. O. and BASSLER, R. S. 1931*a*. Cambrian bivalved Crustacea of the Order Conchostraca. *Proceedings of the US National Museum*, **78** (4), 1–130, pls 1–10.

—— —— 1931*b*. *Indianites*, new name for the Cambrian crustacean *Indiana* Ulrich and Bassler. *Journal of the Washington Academy of Sciences*, **21**, 364.

VANNIER, J. M. C., WILLIAMS, M. and SIVETER, DAVID J. in press. The Cambrian origin of the circulatory systems of crustaceans. *Lethaia.*

WADDINGTON, J., VON BITTER, P. H. and COLLINS, D. 1978. Catalogue of type invertebrate, plant, and trace fossils in the Royal Ontario Museum. *Royal Ontario Museum, Life Sciences Miscellaneous Publications*, 151 pp.

WALCOTT, C. D. 1886. Second contribution to the studies of the Cambrian faunas of North America. *Bulletin of the United States Geological Survey*, **30**, 1–357, pls 1–33.

—— 1887. Fauna of the 'upper Taconic' of Emmons, in Washington County, New York. *American Journal of Science*, **34**, 187–199, pl. 1.

—— 1890. The fauna of the Lower Cambrian or *Olenellus* Zone. *Tenth Annual Report of the United States Geological Survey*, Part 1, 509–763.

—— 1905. Cambrian faunas of China. *Proceedings of the US National Museum*, **29**, 1–106.

—— 1912. Middle Cambrian Branchiopoda, Malacostraca, Trilobita, and Merostomata. *Smithsonian Miscellaneous Collections*, **57**, 145–228.

—— 1913. The Cambrian faunas of China. *Publication of the Carnegie Institution of Washington*, **54**, 1–276, pls 1–24.

WALOSSEK, D. and MÜLLER, K. J. 1992. The 'Alum Shale Window' – contribution of 'Orsten' arthropods to the phylogeny of Crustacea. *Acta Zoologica (Stockholm)*, **73**, 305–312.

WHITE, C. E., BARR, S. M., BEVIER, M. L. and KAMO, S. 1994. A revised interpretation of Cambrian and Ordovician rocks in the Bourinot belt of central Cape Breton Island, Nova Scotia. *Atlantic Geology*, **30**, 123–142.

WHITFIELD, R. P. 1884. Notice of some new species of primordial fossils in the collections of the Museum, and some corrections of previously described species. *Bulletin of the American Museum of Natural History*, **1** (5) (for 1881–1886), 139–154, pls 14–15.

WILLIAMS, M. and SIVETER, DAVID J. 1996. Lithofacies-influenced ostracod associations in the middle Ordovician Bromide Formation, Oklahoma, USA. *Journal of Micropalaeontology*, **15**, 69–81.

—— —— in press. British Cambrian and Tremadoc Bradoriida and Phosphatocopida (Arthropoda). *Monograph of the Palaeontographical Society.*

—— —— HINZ-SCHALLREUTER, I. and MELNIKOVA, L. 1994*a*. On *Cambria sibirica* Neckaja and Ivanova. *A Stereo-Atlas of Ostracod Shells*, **21**, 9–12.

—— —— BERG-MADSEN, V. and HINZ-SCHALLREUTER, I. 1994*b*. On *Vestrogothia longispinosa* Kozur. *A Stereo-Atlas of Ostracod Shells*, **21**, 21–26.

—— —— and PEEL, J. S. 1996. *Isoxys* (Arthropoda) from the early Cambrian 'Sirius Passet' Lagerstätte, North Greenland. *Journal of Paleontology*, **70**, 947–954.

—— —— RUSHTON, A. W. A. and BERG-MADSEN, V. 1994*c*. The Upper Cambrian bradoriid ostracod *Cyclotron lapworthi* is a hesslandonid. *Transactions of the Royal Society of Edinburgh: Earth Sciences*, **85**, 123–130.

WIMAN, C. 1905. Studien über das Nordbaltische Silurgebiet. I. Olenellussandstein, Obolussandstein und Ceratopygeschiefer. *Bulletin of the Geological Institutions of the University of Uppsala*, **6** (for 1902–1903), 36–76, pls 1–4.

ZHANG WENTANG 1974. [Bradoriida.] 107–111. *In* NANJING INSTITUTE OF GEOLOGY AND PALAEONTOLOGY (ed.). [*Handbook of stratigraphy and palaeontology of southwest China.*] Science Press, Beijing, 454 pp., 202 pls. [In Chinese].

ZHANG XIGUANG 1987. Moult stages and dimorphism of Early Cambrian bradoriids from Sichuan, Henan, China. *Alcheringa*, **11**, 1–19.

DAVID J. SIVETER

Department of Geology
University of Leicester
University Road
Leicester LE1 7RH

MARK WILLIAMS

British Geological Survey
Keyworth
Nottingham NG12 5GG

Typescript received 18 January 1996
Revised typescript received 10 September 1996